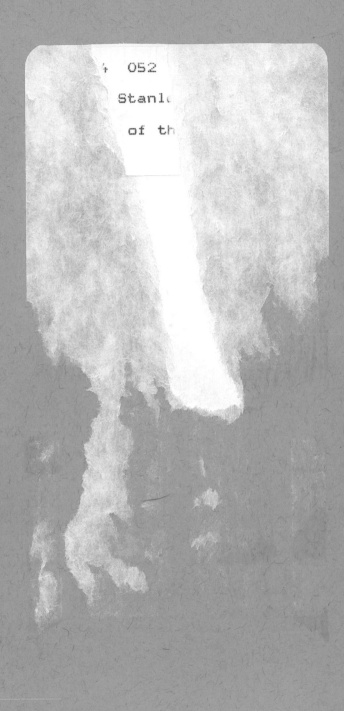

4 052

Stanl

of th

ORIGINS
OF THE
DOMESTIC DOG

Burial site of a prehistoric domestic dog, *Canis familiaris,* from archaeological excavations at Pueblo Bonita Ruin, New Mexico. Photo by O. C. Havens, National Geographic Society.

ORIGINS OF THE DOMESTIC DOG

THE FOSSIL RECORD

Stanley J. Olsen

THE UNIVERSITY OF ARIZONA PRESS

TUCSON, ARIZONA

About the Author

STANLEY J. OLSEN has been a pioneer in the study of mammals associated with Man at archaeological sites. A specialist in zooarchaeology and vertebrate paleontology for nearly forty years, he has conducted extraordinarily thorough research on the osteology of canids. The results of his scientific research have been published in numerous scholarly journals and monographs. He taught at Florida State University for many years and joined the Department of Anthropology at the University of Arizona in 1973, at which time he also took on the duties of Zooarchaeologist at the Arizona State Museum in Tucson.

Unless otherwise noted, all photographs
in this book were taken by Eleanor L. Olsen.

THE UNIVERSITY OF ARIZONA PRESS

Copyright © 1985
The Arizona Board of Regents
All Rights Reserved

This book was set in Compugraphic 8400 Garamond Old Style #3
Manufactured in the U.S.A.

Library of Congress Cataloging in Publication Data

Olsen, Stanley John, 1919–
  Origins of the domestic dog.

  Bibliography: p.
  Includes index.
  1. Canis, Fossil.   2. Dogs—Origin.   I. Title.
QE882. C15045   1985   569'.74   85-1024

ISBN 0-8165-0909-3

CONTENTS

ILLUSTRATIONS

The northern husky *(right)* maintains many of the ancestral characteristics of the wolf *(left)*. However, its muzzle is proportionally shorter than that of the wolf, and the mottled coat is a domestic characteristic. *Drawing by Cynthia Lindquist.*

PREFACE

In the 1980s there has been an increasing interest in the beginnings of animal domestication on a world-wide basis. This interest may stem, in part, from our continuing search for evidence of early human development, particularly that which relates to the beginnings of sedentary communities that are associated with the origins of agriculture, including the domestication of animals.

A particular human interest in the dog *(Canis familiaris)* as the first domesticated animal may be due partly to our long and favorable relationship with this animal as one aspect of many normal family units. In addition, it is the only domestic animal that may be incorporated easily into a hunter-gatherer society. Studies of the different breeds of domestic dogs have been published at least as far back as 150 years ago (Smith 1839). However, there has been no single study made that incorporates the fossil ancestry of the canids with Paleolithic and Neolithic finds.

The wolf *(Canis lupus)* meets the requirements for the ancestral stock from which the domestic dog has risen. This wolf/dog association has been referred to by a number of writers—most recently by Honacki, et al. (1982) in their taxonomic classification of mammals. They state that *"Canis lupus* [is] probably ancestor of and conspecific with the domestic dog, *familiaris."* The wolf has perfected hunting habits that would not have been lost on early human hunters who observed their game-getting practices. Both wolves and humans were pack, or team, hunters early on, and it is likely that these similar tactics for obtaining prey were influential in bringing about the initial stages of association between the two species. This hunting association and the fact that both hominids and large canids have mutually compatable social organizations eventually led to taming and, ultimately, to domestication.

There has been very little published relating to osteological variation observed in normal populations of wild canids. However, a number of archaeological finds that established and pushed back the dates of the "first domestication" of dogs included only fragments of bones and teeth (Lawrence 1968; Turnbull and Reed 1974). In most instances, it could not be stated with certainty that these recorded occurrences were, in fact, *Canis familiaris* and not merely aberrant examples of a local, wild species of *Canis*. Thus, most of the very early dates for dog domestication must be questioned and set aside until material complete enough for observing and comparing multiple morphological characters has been obtained. Such comparisons are required for a more positive determination as to whether *Canis familiaris* is present or absent in a particular archaeological context.

One of the problems of determining whether an archaeological find represents an early domestic dog or a local wild species of the same genus is the close osteological resemblance of most similar-sized species of *Canis* to each other. Among the earlier paleontologists who outlined and presented a phylogeny of dogs was W. D. Matthew, who published a basic guide many years ago (Matthew 1930). Although much of the discussion that follows was based on numerous new finds that were made subsequent to Matthew's phylogeny, his work still provides a solid footing for discussing origins of the dog.

No in-depth discussion of canid evolution or the processes of early dog domestication can be undertaken without considering the many finds made during the 1970s and early 1980s in China. It was only during that time that, due to increased communication between scholars in the People's Republic of China and the United States of America, a regular exchange of ideas among vertebrate specialists in these two nations came about.

The critical collections in the People's Republic of China were examined between 1976 and 1984. Pertinent specimens in the Institute of Vertebrate Paleontology and Paleoanthropology and the Institute of Archaeology in Beijing were measured and photographed. The same was true for canid material in the Tianjin Natural History Museum and in the Museum of Inner Mongolia in Huhehot.

It should be noted that, as new discoveries are made and new methods of analysis are developed, some of the conclusions set forth here will be altered, just as Allen's (1920) monograph on aboriginal dogs was updated. The many volumes relating to wolves and dogs that have appeared all share one discrepancy: they do not illustrate any of the canid skulls which show the diagnostic morphological changes that are used as evidence to distinguish one group of canids from another. This is particularly true for the fossil forms. I have attempted to eliminate this deficiency and, at the same time, present illustrations of specimens that have not been shown before.

The entire picture of dog development is far from complete, and there will be many gaps in any presentation. It is hoped that this book will be a base upon which future canid finds may be evaluated so that eventually we will have a more continuous picture of what has occurred in dog evolution from the late Paleolithic to comparatively recent times.

ACKNOWLEDGMENTS

First and foremost, I want to acknowledge and credit my son, John W. Olsen, with the painstaking task of compiling the section on canid domestication in Asia (particularly China) and for critically editing the manuscript. Much of this work required translating many important articles written in Chinese so that the data could be included here. I also wish to acknowledge the kindness and assistance of my colleagues in the People's Republic of China, particularly Wu Ju-kang, Chou Ming-chen, Qi Guo-qin, Wu Shin-zhi, and Li You-heng of the Institute of Vertebrate Paleontology and Paleoanthropology in Beijing. Xia Nai, An Zhi-min, and Zhou Ben-shun made it possible to compare and photograph Neolithic dogs from China and Inner Mongolia in the collections under their care at the Archaeological Institute in Beijing. Wen Hao, Director of the Museum of Inner Mongolia in Huhehot, was most generous in allowing access to the vertebrate collections there. I also thank Li Rong of that museum for allowing me access to pertinent specimens and for his assistance on my visits to Huhehot, Inner Mongolia. I extend my deep appreciation to Huang Wei Long of the Tianjin Natural History Museum for allowing me full access to the Pleistocene mammal collections there. Thanks are due to R. Tedford and Beryl Taylor of the Department of Vertebrate Paleontology of the American Museum of Natural History in New York City and to Sidney Anderson and Richard Van Gelder of the Mammal Department of the same museum. They allowed complete access to the canid collections under their care. Barbara Lawrence and Edith Rutzmoser of the Museum of Comparative Zoology at Harvard University, were, as always, most generous with their assistance. Most of the illustrations are due to the careful work of Eleanor L. Olsen, who photographed specimens in the U.S., China, and Great Britain. Juliet Clutton-Brock kindly allowed me to examine and photograph the important canid skull from the Star Carr site. Casts of pertinent canid material were graciously furnished by William Turnbull of the Field Museum of Natural History in Chicago, Brenda Beebe of the

Department of Anthropology at the University of Toronto, R. L. Reynolds of the George Page Museum at Rancho La Brea in Los Angeles, and Gerald Schultz of the Department of Geology at West Texas University, Canyon, Texas. I am also grateful for the use of the Bagnell site skulls, which were lent to me by Donald Lehmer of Dana College, Blair, Nebraska.

I am deeply appreciative of the difficult task undertaken by Doris Sample in order to bring the rough manuscript into a finished form. I wish to gratefully acknowledge the generous support of Ann Fallon in the publication of this study. I also want to thank Edwin H. Colbert for reading the manuscript and for offering helpful suggestions. Any omissions or interpretations of published literature are entirely my own. Research in China was aided by a grant from the National Geographic Society.

STANLEY J. OLSEN

1. THE FOSSIL ANCESTRY OF *CANIS*

If one were to draft a phylogenetic tree depicting the origins and relationships of canids from their earliest fossil record to the present, it would, of necessity, be a tree of many branches. Most of the branches would fall short of the length that we would like to see. Some would require dotted breaks between the main trunk and their known limits. This incomplete structure would be due primarily to our lack of an adequate fossil record at the beginnings of canid ancestry.

Fossils represent only a very small percentage of all the animals that have lived and died. Nearly ideal taphonomic circumstances must prevail from the time of an animal's death, through the process of interment, continuing on through the mineral replacement of bone cells, to eventual fossilization (Behrensmeyer and Hill 1980; Brain 1981). This entire process may take millions of years, which add to the possibility of destruction of the remains by weathering and through other agencies. The recovery of skeletons is also, to a great degree, dependent upon chance. The geologic beds where such bones may be buried must be of a nature that allows them to be located and collected for study. Therefore, search areas are, for the most part, limited to arid or semi-arid lands, where there is little in the way of covering vegetation to hide the remains from detection and where there is considerable erosion to remove the covering soil and leave them exposed. Obviously, fossilized animals entombed in regions that are covered by heavy vegetation have little chance of being discovered and contribute nothing to our knowledge of these early forms.

All animals are not treated equally by nature in the process of entrapment and preservation. Many carnivores, due to their agility, are often able to extricate themselves from natural traps, such as quicksand, cave-ins, or

1

sinkholes, that would seal the fate of many heavier-bodied, less agile herbivores, such as proboscideans, horses, camels, or bovids. However, enough fossil carnivore material has been collected and studied by vertebrate paleontologists to present us with an acceptable idea of the ancestry of canids leading from the early Tertiary forms to the present domestic dog.

Much of fossil mammal taxonomy is based on animal dentitions. This interpretation is due, in part, to the fact that teeth are among the most durable elements of mammalian skeletons and are preserved long after various taphonomic factors have caused the disintegration of many softer parts of an animal's skeleton. The overall dental morphology and structure indicate the animal's diet and are, therefore, useful in arriving at the correct taxonomic classification of a fossil species.

One of the problems of determining whether an archaeological find represents an early dog or a local wild species of the same genus is the close resemblance of the dentitions of most similar-sized recent species of *Canis* to one another. Even when comparing fossil canids with recent forms, it is apparent that the dentitions of present-day canids have not changed drastically from those of their fossil ancestors that date back to the Miocene epoch at least 15 million years ago.

The earliest fossil carnivores that can be linked with any certainty to the canids are the miacids, whose presence can be traced back to the transitional upper Eocene-lower Oligocene epoch (about 40 million years ago). Certain osteological features of the crania of these early fossil mammals are of taxonomic importance in suggesting their canid affiliation, but they do not allow for the determination of a close phylogenetic relationship. The lack of fully ossified bullae is one characteristic of miacid fossils. This area of the skull is diagnostically distinct among various groups of carnivores and is important in arriving at broad taxonomic determinations within this group of animals.

The miacids were small carnivores—the largest was about the size of a fox, the smallest were ferret-sized. Their bodies were long, with shortened limbs, plantigrade feet, and a rather long, extended tail. The miacid dentitions already possessed the characteristic prominent carnassials so typical of the later carnivores.

At this early period of carnivore radiation, it is contradictory, but understandable, to refer to these various and differing ancestral animals as cat-like dogs, mustelid-like dogs, hyaena-like dogs, or bear-like dogs. Such descriptions have been used to describe animals that have some features suggesting carnivore characteristics. However, these descriptions can only be broadly applied to these earliest carnivores. Eventually these carnivores died out, leaving in this particular line of evolution only the dog-like dogs that led to *Canis*.

Dogs and foxes originated in the Western Hemisphere; their apparent beginnings were in the Oligocene through the Miocene epochs (some 15 million years ago), during which time they extended their range to the Old World. In this cross-continental migration they joined the camel and horse as animals that originated in North America and eventually migrated to other continents. The earliest recognized animal that is in the direct line of dog evolution is the small, fox-like animal *Hesperocyon* (formerly *Pseudo-cyonoides*). Their known skeletons indicate that they had a fox-like appearance in life, with long, muscular bodies and long tails. Their feet, which were digitigrade rather than plantigrade, allowed the animals to walk on their toes, like modern dogs, rather than flat-footed, like the early miacids.

A descendent form that developed in the late Oligocene epoch and died out in the early Miocene was *Mesocyon*. This animal had shearing carnassials and a rather massively proportioned skull with a short muzzle. Although it was about the size of a coyote (*Canis latrans*), its appearance and its habits, if we are to judge from its dentitions, were more like those of a hyaena than a wolf or a dog. In the 1960s and 1970s some research workers concerned with canid ancestry regarded the Miocene animal *Tomarctus* as a basic ancestral canid (Fox 1971; Hall 1978; Mech 1970; Repenning 1967; and others). Other workers believed that *Tomarctus* was more properly in line with the borophagines, a group of carnivores regarded as hyaena-like in form and habits (Matthew 1924; Merriam 1906; Vanderhoof and Gregory 1940).

One particular characteristic of the skull that supported the latter theory is in the area of the rostrum and muzzle (Figs. 1.1, 1.2). In *Tomarctus* the joining of the frontals and the posterior extensions of the premaxillae prevent the nasals from touching the maxillae. This condition is also apparent in the borophagines and in the hyaenas. It is not present in members of the canini. The purpose of these peculiar bone extensions may be related to the nipping bite that is apparently important in the feeding and fighting habits of this group of carnivores. The teeth that are most important in this action are the incisors. It seems reasonable to assume that, in order to counteract the stress placed on the anterior end of the muzzle, the long processes of the premaxillae extended back to corresponding processes of the frontals to form a strong buttress.

This theory is not new. It was expressed by Merriam (1906) in his discussion of the skull of a new genus of carnivore, *Tephrocyon*, from the Miocene John Day beds in Oregon. *Tephrocyon* was later placed in the genus

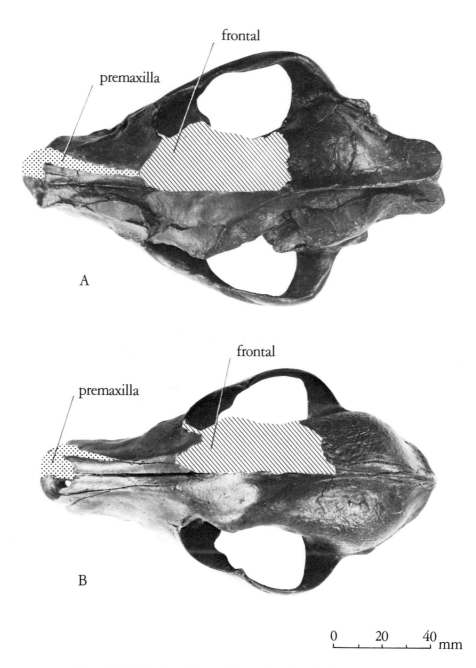

premaxilla

frontal

A

premaxilla

frontal

B

0   20   40
|—————|—————|———— mm

Figure 1.1. Relation of frontal bone (cross-hatched) to premax-
illae (stippled) in *Tomarctus* and domestic dog: A)
Miocene *Tomarctus* dorsal aspect of skull; B) recent
Canis, dorsal aspect of skull.

4

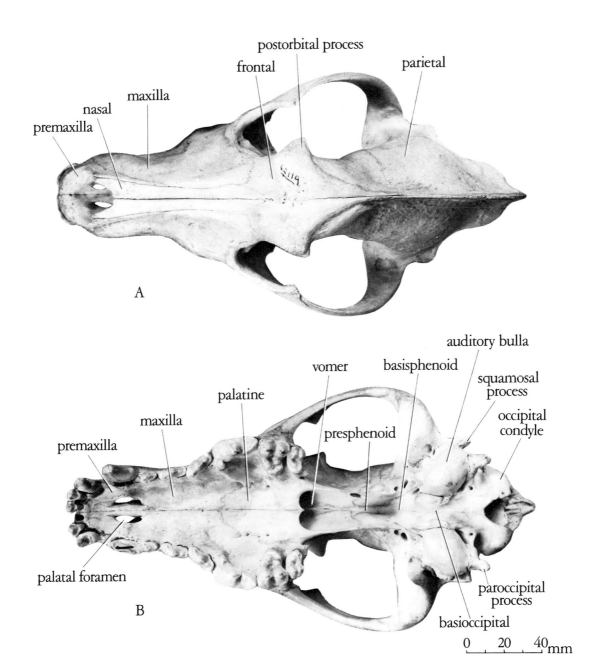

Figure 1.2. Elements comprising the canid skull: A) dorsal aspect of skull of a modern wolf, *Canis lupus;* B) palatal aspect of same skull.

Tomarctus by Matthew (1924) based on new material from the Snake Creek Miocene beds in Nebraska. Matthew also noted the peculiar extensions of the premaxillae and frontals. Tedford (1978), reviewing the ancestry and taxonomic affiliations of *Canis*, also determined that this characteristic of *Tomarctus* was not found in any members of the canini.

The upper carnassials and molars of *Tomarctus* are proportionately heavier in construction than in canids, and they approach a square occlusal surface, unlike those found in typical canids. This square shape suggests a decided crushing action for these teeth similar to the later hyaena-like borophagines. The high, thin saggital and occipital crests and the blunt, foreshortened muzzle again suggest hyaenid characteristics. Thus, it doesn't appear that *Tomarctus* was in a direct line of canid evolution, but, rather, that it branched off in the direction of the hyaena-like borophagines.

By contrast, the anterior teeth of canid dentitions are ideally suited for ripping and tearing flesh and the posterior molars are only moderately designed for crushing (Fig. 1.3). At the anterior end of the upper and lower jaws are located sharp, chisel-edged incisors for nipping and pulling tough tissue from the bone, followed by more or less large, strong canines for piercing—necessary for killing prey or for defense. Premolars, with sharp, high-crowned cutting edges, are located behind the canines and are well designed for cutting flesh and cartilage. The largest of these in dogs follow the smaller premolars and are adapted for shearing tendons as well as flesh and some softer bone. These teeth, one upper and one lower on each side, are the carnassials and are the most specialized teeth in carnivores. In dogs they are the upper fourth premolar and the first lower molar. The opposing carnassials, which occlude in chewing, are similar in action to a pair of shears, cutting much that is placed between them. The remaining teeth behind the carnassials, as well as the heel of the carnassials, have low, multiple-cusped crushing surfaces for breaking bones. This arrangement completes an ideally designed dental mechanism for an animal that is basically a meat eater.

One late Oligocene carnivore, *Leptocyon*, possesses skeletal characteristics that make it a candidate for the ancestral canid. It was recognizably fox-like in body form and dentition. Leidy (1858) briefly described this new species under the name *Canis vafer* and stated that it was decidedly like the living red fox (*Vulpes fulva*). Matthew (1918) believed that this animal was not directly related to *Canis* and erected the new genus *Leptocyon* to include the species *vafer*. He felt that this species paralleled the foxes but did not suggest direct affinities to modern *Canis*.

I agree with Tedford (1978) that *Leptocyon* is in a direct ancestral line leading from the Miocene forms to the more modern *Canis*. However, rather

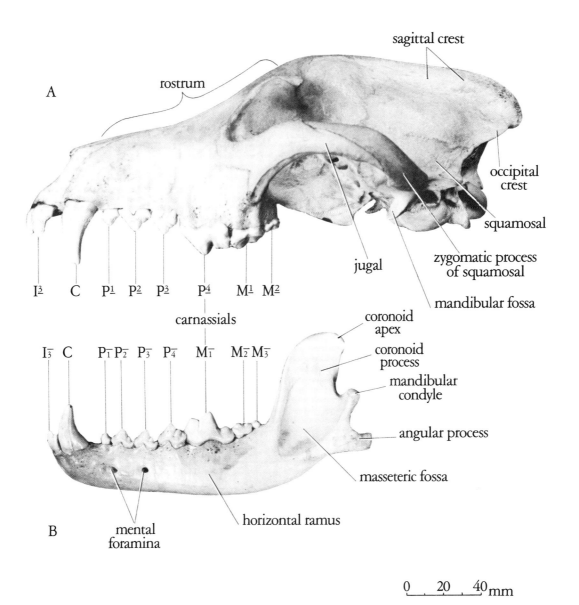

A

rostrum

sagittal crest

occipital crest

squamosal

zygomatic process of squamosal

jugal

mandibular fossa

I^3 C P^1 P^2 P^3 P^4 M^1 M^2

carnassials

coronoid apex

coronoid process

mandibular condyle

I_3 C P_1 P_2 P_3 P_4 M_1 M_2 M_3

angular process

masseteric fossa

horizontal ramus

B   mental foramina

0   20   40 mm

Figure 1.3. Elements comprising the canid skull and mandible: A) left lateral aspect of skull of modern wolf, *Canis lupis;* B) lateral aspect of left mandible of modern wolf, *Canis lupus.*

than to indulge in taxonomic semantics, I accept the present generic assignment of *Leptocyon*, rather than *Canis*, for the time being. Determination of the stages of evolution from *Leptocyon* to *Canis* is, for the most part, based on visual observations made by the many individuals who have been concerned with this problem. These determinations are as varied as the scientific backgrounds of the people who made the comparisons.

If the person conducting the analysis believes in taxonomic splitting, using minute morphological criteria as the basis for establishing whether the animal in question belongs in an already established genus or species or in a newly designated taxon, then there will be numerous genera and species recorded. On the other hand, if the analyst believes in lumping forms into fewer genera or species, based on broader morphological characteristics, then fewer genera and species will be recorded.

Vertebrate paleontologists have often been too eager to establish a new classification for an animal based on inadequate material. More often than not the simple fact that skeletal remains were found in a new locality and from a geologic horizon that was a bit older or younger than a previously described species was reason enough to erect a new genus or species for a new find.

At times, even complete skulls and skeletons of individual animals were assigned to new specific taxonomic categories simply because they were from older Pleistocene deposits. For example, the coyote from Rancho La Brea, California, was assigned the new name *Canis orcutti* (Merriam 1910), although the living species and subspecies in that area is *Canis latrans ochropus* (Hall and Kelson 1959). The same is true for the coyote from the Papago Springs Pleistocene cave deposits in Arizona, which was assigned the name *Canis caneloensis* (Skinner 1942). The living coyote of the area is *Canis latrans mearnsi* (Hall and Kelson 1959). Both of these "extinct" canids are within acceptable limits of individual variation within the living species. Nowak (1979) put both in synonymy with the living coyote (*Canis latrans*), but he allowed the specific identifications that were assigned to the fossil forms to remain as subspecific designations. With this I agree.

Occasionally, unfamiliarity with standard, accepted taxonomic rules and procedures accounts for erroneous interpretations which creep into the published literature. For example, Singh (1974:55) interprets the indentification *Canis* sp. from a Levantine Neolithic faunal list as "dog species (*Canis* sp.)." In reality, *Canis* sp. indicates that the genus *Canis* is present in that site, but that the species could be either wolf, jackal, or domestic dog (Schenk and McMasters 1956). Once this determination is mistakenly carried over as dog, a future writer may refer to it as occurring there without question.

8

Range of morphological variation of the genus or species under consideration is, at times, not considered in the apparent rush to establish a new genus or species or even higher taxonomic category. At times, assigning a new name to a small fragment of jaw or a single tooth was based on individual judgment alone, rather than on quantitative considerations. A great many of these errors in classification were made within the genus *Canis* as well as in establishing the earliest geological (archaeological) presence of *Canis*. Many of these determinations still stand unchallenged in the literature (Turnbull and Reed 1974; Galbreath 1938). A few have been synonymized with previously known and accepted species, where they rightfully belong (Nowak 1979).

What degree of slight morphological variation is acceptable in a visual analysis that establishes a canid genus or species if only fragmentary bones or teeth are present? The answer can, in all probability, never be specifically stated, since the interpretation will be based on the powers of observation and experience of the person who is recording his or her observations and comparisons.

In the discussions relating to the "first" occurrence of *Canis* in the fossil record, I refer to basic forms that are generally accepted by the scientific community that is concerned with this problem and do not review the earlier-reported occurrences of *Canis* that are currently considered synonyms of other genera or species. As more data become available, and as new paleontological finds are made, many of the older described finds of *Canis* will no doubt be subject to changes in classification.

To attempt a particular affinity of the earliest-recorded *Canis* with a living member of this genus would be stating more than is known. Johnston (1938) lists his type species of *Canis lepophagus* from the Pliocene beds of Cita Canyon, Texas (Fig. 1.4) as "an ancestral coyote." However, he does not present convincing evidence that there are any morphological features that separate the ancestral coyote from the ancestral wolf at this early stage of evolution of *Canis*. Kurten and Anderson (1980:167) believe, as I do, that *Canis lepophagus* may be an ancestral wolf. I believe that it is sufficient at present to state that Johnston's (1938), Björk's (1970), and Hibbard's (1941) species of *Canis* from the Pliocene are simply ancestral or early species of *Canis*, rather than to attempt to associate them directly with either the coyote (*Canis latrans*) or the wolf (*Canis lupus*), both of which are known from a number of late Pleistocene deposits in North America.

Hibbard (1941), in his discussion of *Canis lepophagus* from the Rexroad Pliocene fauna of Kansas, identifies a right, lower, second molar from these same beds as belonging to *Canis* sp. He states that it represents "a dog as large as *Canis latrans*. Size alone distinguishes it from *Canis lepophagus*."

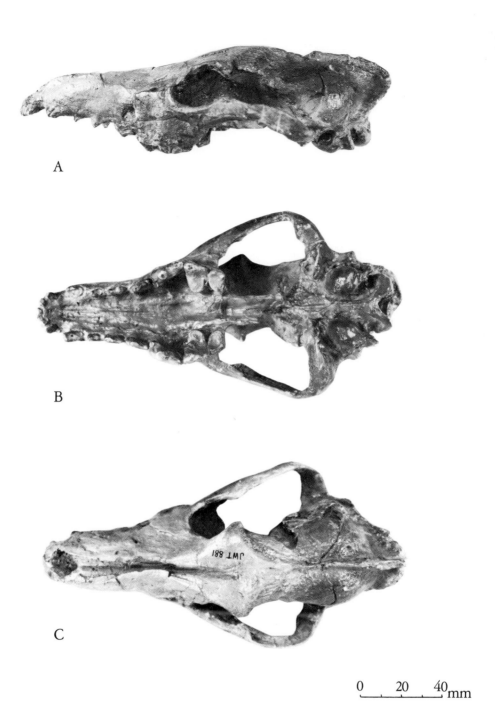

Figure 1.4. Skull of *Canis lepophagus* (earliest known species of genus *Canis*) from Pliocene beds of Cita Canyon, Texas: A) left lateral aspect of type skull; B) palatal aspect of type skull; C) dorsal aspect of type skull.

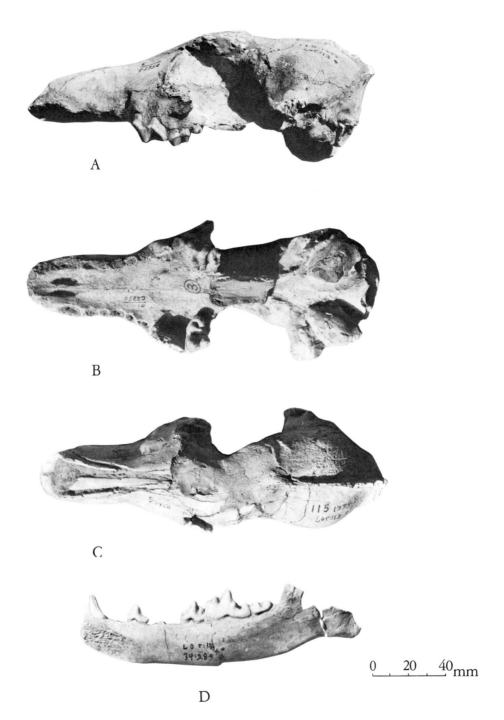

Figure 1.5. Skull and mandible of late Pleistocene wolf, *Canis lupus variabilis,* from Zhoukoudian, China: A) left lateral aspect of skull; B) palatal aspect of skull; C) dorsal aspect of skull; D) medial aspect of right mandible.

11

Needless to say, the use of "dog" in this context is incorrect and misleading. Given the age (Pliocene) of the fossil, it can in no way be considered a dog, unless one refers to all members of the genus *Canis* as dogs. Referring to a single tooth of a canid as being diagnostic for identifying a species of *Canis* is not a valid determination. Individual morphological variations alone could bring about erroneous interpretations. The hypodigm of *Canis lepophagus* is so small that no accurate characteristics or individual variation can even be inferred, let alone establish with any certainty whether *lepophagus* is indeed a valid species.

We know that wolves (*Canis lupus*) and coyotes (*Canis latrans*) can cross and produce hybrids and that both of these animals can mate with the domestic dog (*Canis familiaris*) with similar results. On the other hand, even though mating may be possible between the wolf and coyote, there are many social factors, as well as size differences between average individuals of these two closely related wild species, that would preclude such unions as being common occurrences, even in those areas where the wolf and coyote are sympatric.

A karyological approach, detailing the chromosome number and morphology of the various species of *Canis*, gives us a rather reliable insight into the taxonomic associations of the members of this genus. The chromosomes play a role in establishing the taxonomic affinities of the closely related species of *Canis*. In the wolf, coyote, jackal, and domestic dog the diploid number (78) and the so-called fundamental number (80) are the same (Fox 1975). These data vary considerably for many of the other canid species.

The proponents of the theory that the jackal (*Canis aureus*) and the coyote (*Canis latrans*) were the ancestral stock of the domestic dog (*Canis familiaris*), regardless of geographic location, must take into consideration that neither of these wild species occur in China, where small domestic dogs occur as far back as the early Neolithic (ca. 5000 B.C.). However, a small wolf (*Canis lupus variabilis*) has been found in a number of early to late Pleistocene localities in China (Fig. 1.5) in association with both *Homo erectus* and *Homo sapiens*. The taxonomic position of these Chinese Pleistocene wolves, as well as their relationship to early dogs in China, is discussed in Chapter 4.

In those areas outside of Asia where Paleolithic and Neolithic dog remains occur, most of the wolves, both living and fossil, are of a size considerably larger than the largest domestic dog remains. The question then arises as to what form or forms of canid bridged this gap between the two differing size groups. This is not to say that the larger domestic dogs found in a prehistoric site in the northern part of North America could not have been derived from the local populations of large wolves. In fact, there is considerable evidence that points to this derivation as a distinct possibility. The wolf,

Canis lupus, has a well-established record in North America dating back to the early Pleistocene. The absolute size and proportions of the skull and dentition of the gray wolf *Canis lupus* set it apart from all other canids within its range (except the dire wolf), and these characteristics have, for the most part, remained virtually unchanged from the middle Pleistocene to the present. This wolf has been described as occurring in a number of sites of Pleistocene age throughout North America (Brown 1908; Colbert 1950; Merriam 1910; and others).

A small subspecies of wolf, *Canis lupus edwardii* (Gazin 1942), known from the early Pleistocene of the southwestern United States, resembled the small, recent, red wolf, *Canis lupus rufus*. This smaller, early Pleistocene wolf—and possibly some other small, primitive wolves—entered Eurasia by way of the Bering Strait, where it (or they) may have given rise to the small, Chinese Pleistocene wolf, *Canis lupus variabilis*, and thence to the modern form of Chinese wolf, *Canis lupus chanco*.

There is a theory that the Eurasian gray wolf moved back across the Bering Strait as early as the Kansan glaciation of the Pleistocene to once more occupy North America. The supporting osteological data is too incomplete to either support or disprove this theory.

A statement must be made with regard to the largest of the Pleistocene wolves—the dire wolf, *Canis dirus*—even though it is not to be considered in any way as a likely progenitor of the domestic dog. The dire wolf is known particularly from the Rancho La Brea tar pits of California, although it has been found elsewhere in the United States, including Florida. It was a highly specialized species and was not ancestral to any of the modern species of wolves. The large size of this animal distinguishes and separates it from all other members of the genus *Canis*. There is no overlap in size with other canids when dire wolf remains are subjected to a multivariate analysis.

When describing and assigning new taxonomic names to animals or when reclassifying already named forms, the rules set forth by the International Commission on Zoological Nomenclature should be followed. The purpose of following this procedure is to bring order into systematic biology and to avoid having new classifications in print that are dependent on the personal opinions of various writers. Sadly, the procedure is not always followed. Nearly every publication that discusses the systematics of classification presents columns of previously described species that are then listed as synonyms of the particular species name chosen by the writer (Nowak 1979; Mech 1974).

Writers often select an animal rightfully listed as a subspecies and advance it to specific rank. For example, the red wolf, *Canis lupus rufus*, is generally listed as *Canis rufus*, but, actually, the separating characteristics are no greater in value than those generally used to establish and separate subspecies of other canids. Nowak (1979) recognized this problem when he stated that ". . . it is sometimes difficult to separate specimens of red and gray wolves."

I believe that it would serve no purpose to add long lists of taxonomic changes for and synonyms of *Canis lupus*, or to list animals named on single teeth or on lost specimens. Rather, it is of more value to outline the mainstream of canid evolution that led from the miacids and culminated in the wolves and, eventually, the domestic dog.

2. TAMING AND DOMESTICATION OF EARLY WOLVES

There is adequate evidence from paleontological sites in North China indicating an association of hominids (*Homo erectus pekinensis*) and small wolves (*Canis lupus variabilis*) from as early as the middle Pleistocene. The association of *Canis* with *Homo*, particularly at Zhoukoudian, implies no more than that they were contemporary animals at a very early stage in the development of both animals. It places them in close proximity and allows for this small wolf to evolve into a tame and domesticated animal at a much later state with *Homo sapiens*. No taming or early domestication of *Canis* is inferred at this early association of both the wolf and hominids at Zhoukoudian.

Studies relating to the social structures of both humans and wolves shed some light on what *may* have taken place in the initial development and relationship of these two groups of early hunters that eventually culminated in the close relationship between dogs and humans that prevails today throughout much of the world. Foremost among these studies are those by Mech (1970), Fox (1971), and López (1978). Perhaps the best compilation that relates to this relationship is a series of papers edited by Hall and Sharp (1978). Several theories dealing with the attitudes of humans toward wolves and of wolves toward humans are discussed in detail in their volume.

Early human hunter-gatherer societies and wolf packs are similar in a number of respects. Both are comprised of social units that are relatively small in number. Both are capable of hunting over open ground or wooded areas, pursuing rather large game and exerting considerable physical effort and energy over prolonged periods to accomplish their goals. Both use hunting methods that require a pack, or team, effort rather than that of a lone hunter.

Faunal evidence from paleoanthropological sites of late Pleistocene age (ca. 10,000 B.P.) indicates that *Homo sapiens*, at this stage of development, depended on a diet that included a sizeable portion of game animals. The wolves of this period, being carnivores, also depended on wild game for their food source. The hunting and foraging groups of both species would be, of necessity, rather small. Since hunting game by running it down takes a good deal more time than consuming the kill, one of the characteristics of both these predator groups would be sharing the spoils of the hunt. The resulting socialization attached to this group process would be developed independently by both wolves and humans (Hall and Sharp 1978).

A social feature shared by wolves and humans is that the labor required for a hunting existence does not follow strict sexual lines; rather it is based on the individual's maturity and ability to hunt successfully. The non-hunters of the pack (immature animals), therefore, would gather in rendezvous areas that were selected for regrouping and for the sharing of freshly killed game provided by the active hunting members of the pack (Hall and Sharp 1978; López 1978; Mech 1970). The wolf pack's methods of obtaining game appear to be well organized and planned and should be briefly abstracted here as a possible model of how human hunting societies may have obtained game during the Pleistocene.

As pack hunters wolves are able to capture and kill animals much larger than themselves. Artiodactyls such as deer, mountain sheep, elk, caribou, and even moose, a formidable adversary, are all chosen prey. Rabbits, beaver, and, at times, fish in shallow water, as well as other small mammals are also caught and eaten (Mech 1970). Wolves have been observed to change their hunting tactics more to suit the terrain and weather conditions than to suit the habits of a particular animal that they are pursuing. Their ability to procure food is not always successful on a regular basis; rather, feast or famine may be more the norm for these animals. The selected prey in an ungulate herd may be an old or weakened animal, but this is not necessarily the rule. Healthy, prime adults, as well as immature animals, are also selected and killed by wolves (Mech 1970). The method of bringing down such an animal is for the wolf to run alongside, slashing and tearing at its hindquarters, flanks, and abdomen until it is weak enough from its injuries to be dispatched (López 1978). Wolves may grab an animal, such as the moose or caribou, by the nose, holding on while other members of the pack close in for the kill (Hall and Sharp 1978; López 1978; Mech 1970). Wolves are intelligent hunters, and, whenever possible, they choose animals that are easier to procure than healthy, active prey. Animals that are injured, infected by parasites to such a degree that they are weakened, or foundering in deep snow are all known to have been selected by wolf packs for food (López 1978). Animals in this condition, of course, signal their plight to some degree; these signals would not be lost on keen observers, such as wolves (or humans), and the weakened prey would be relatively easy to dispatch.

If a prey animal runs, it is certain to be pursued, especially if it is a lone individual (Mech 1970). To weaken such an animal, particularly if it appears to the wolf to be a strong, healthy adult, the practice of relay hunting may be put into practice, taking advantage of the terrain to wind the pursued. It has been observed that a wolf pack of four to six animals may send out two wolves to herd a lone victim into an ambush of the remaining members of the group (Mech 1970). The hunt may end in a rush, terminating the animal's struggles in a few moments, or the dogged pursuit may go on for miles before the pack closes in.

Wolves that have maintained a territory for a considerable length of time may know the terrain well enough to use shortcuts to intercept running prey. They are also known to hunt into the wind when approaching a feeding herd of artiodactyls.

We can observe wolves hunting in their natural habitat and make the assumption that their methods are little changed from those employed by similar, if not taxonomically identical, animals during the Pleistocene. We cannot, with the same assurance, assume what hunting methods were used by humans during this same period. It is possible, however, to obtain some idea of hunter-group size, types of hunting weapons employed, kinds and ages (and at times individual sex) of prey animals, and the season in which these animals were hunted. Such data can often be compiled from paleontological or archaeological excavations, if the preservation is adequate.

Ethnoarchaeological studies of the hunting practices of contemporary hunter-gatherers in Australia, Africa, or in the Arctic area of North America may offer a suggestion as to how their Pleistocene counterparts may have procured game. However, I feel that extrapolating such information in an attempt to reconstruct early wolf-human hunting patterns or even associations would be greatly speculative.

woolly mammoths (*Mammuthus primigenius*). The remains of the wolves that had proportions of the living wolf of that area today (*Canis lupus albus*) were also found, along with the three or four short-faced animals. Pidoplichko coined the taxonomic designation "*Canis lupus domesticus*" for these short-faced wolves. I propose to change the taxonomic assignment to *Canis lupus familiaris* for these and other similarly differing wolves and to use the name designated for the domestic dog to identify this Pleistocene race or subspecies, rather than to erect a new trinomial designation. The use of *familiaris* follows the correct taxonomic progression from a wild wolf, through taming, to the domestic dog. This term seems to be less confusing than the name *domesticus*, which has commonly been used as the scientific name for the domestic cat (*Felis domesticus,* now *F. cattus*).

For a number of years, particularly during the 1930s, the Frick Laboratory of the American Museum of Natural History in New York City had field representatives stationed in the Fairbanks area of Alaska. They were there to cooperate with the University of Alaska and the Fairbanks Exploration Company in order to collect faunal remains as they were uncovered in the muck deposits by hydraulic gold-mining operations. Due to these placer-mining methods, it was impossible to determine the original stratigraphic context of a particular bone or artifact, since many had been collected by the gold-dredge crews from the spoil dumps of reworked and discarded matrix.

There is a general consensus that artifacts from such areas may have originally been on the surface, but might later be sighted several meters below the ground level after having been mixed with the bones of extinct animals (Rainey 1939). It might have been possible to find some artifacts in association with the older, extinct animals if different methods of mining had been practiced or if it had been possible to employ controlled scientific excavations, but these situations were not possible when the bulk of known Pleistocene vertebrate material was collected in the 1930s.

Between the years 1932 and 1953 twenty-eight more or less complete wolf skulls were obtained from the muck deposits in an area north and west of the city of Fairbanks, Alaska. The age of the deposits from which these wolf skulls were obtained has been determined as upper Pleistocene, Wisconsin Age (about 10,000 years ago). The geologist T. L. Péwé published two reports (1975a, 1975b) on the geology and stratigraphy of this area and the associated recovered fauna. The wolf skulls were collected from dredging operations on Ester Creek, Cripple Creek, Engineer Creek, and Little Eldorado Creek. All of these creeks also yielded Paleolithic artifacts (Rainey 1939). However, the association of these artifacts with the skeletal remains of extinct animals is problematical, due to the collecting methods that had to be employed. The

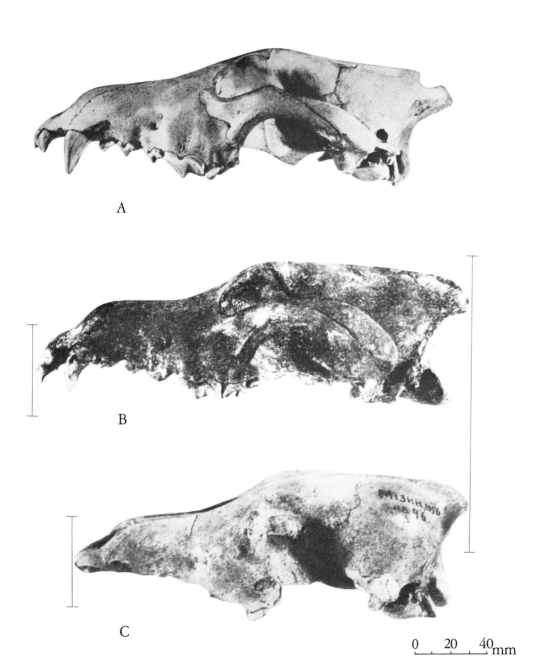

Figure 2.1. Comparison of skulls of present-day wolf, late Paleolithic wolf, and late Paleolithic short-faced wolf: A) left lateral aspect of present-day wolf, *Canis lupus lupus*, from Kiev region, Ukraine; B) left lateral aspect of skull of a late Paleolithic wolf, *Canis lupus*, from mammoth kill site, Mezin, Ukraine; C) left lateral aspect of skull of a late Paleolithic short-faced wolf from mammoth kill site, Mezin, Ukraine. After Pidoplichko (1969).

mechanical mixing of artifacts and bones, referred to above, prevented the collectors from obtaining any possible evidence of contemporary association of Pleistocene wolves and humans in this area of Alaska.

T. Galusha, of the Department of Vertebrate Paleontology at the American Museum of Natural History in New York City, worked for a number of years on the wolves from the Fairbanks dredging operations. He noted that a number of the skulls were extremely short-faced for wild wolves (Figs. 2.2, 2.3) and approached modern Eskimo dogs in facial proportions, although they were considerably larger animals overall. He turned the project over to me shortly before his death. I continued his studies and compared this collection with a large series of both Pleistocene and Holocene wolves as well as with Eskimo dogs from both Greenland (Fig. 2.4) and Siberia (Fig. 2.5). I feel confident at this point in stating that the short-faced wolves from the Fairbanks area appear to be forerunners of the later, domesticated Eskimo dogs.

The proportions of the skulls of these wolves that vary do so in the rostral area. The area of the skull that is anterior to the infraorbital foramen is noticeably foreshortened and constricted laterally in several of the skulls. Two of them (F:AM70933 and F:AM67156) lack anterior premolars. Several others have the full complement of teeth, but the premolar series is set at an oblique angle to accommodate all of the teeth in the abbreviated dental margin. The dentition of F:AM67153 is considerably smaller than the average wolf of that area and approaches the overall tooth size found in the Eskimo dog.

Dishing of the rostrum, when viewed laterally, is evident in all of the short-faced skulls identified as *Canis lupus* from the Fairbanks gold fields. The occipital and supraoccipital crests are noticeably diminished compared to those found in average specimens of *Canis lupus*. The occipital overhang of these crests, a wolf characteristic, is about equal in both groups of *Canis lupus*. Multivariate analysis of all of the wolves from the Fairbanks collection separated the collection into two groups, based on either their predominant wolf-like or dog-like characteristics. The results of this analysis are still, for the most part, speculative regarding the causes for this unique, short-faced condition observed in a small sample of otherwise normal Pleistocene wolf skulls. Examination of a large series of recent wolves from the Alaskan area did not produce individuals with the same variations as those from the Fairbanks gold fields.

Perhaps it will never be possible to establish if there was a close human association with the Alaskan Pleistocene wolves, due to the types of deposits in which both the wolves and the lithic assemblages occur and to the conditions under which both were collected.

In 1939 and 1940 an Eskimo village site was uncovered that dated to the first millennium B.C. This village, named Ipiutak, is believed to represent an occupation of early Eskimo immigrants from Siberia. The artifacts suggest a Neolithic culture (Larsen and Rainey 1948). The site is located on the tip of Point Hope, Alaska, approximately 125 miles north of the Arctic Circle and is at the westernmost point of the continent north of the Bering Strait. At least five more or less complete *Canis* skulls were recovered from the archaeological excavations at Ipiutak. These were studied and reported on by Murie in an appendix to the site report by Larsen and Rainey (1948). After comparing these skulls to those of wolves, Eskimo dogs, and other large domestic dogs, O. J. Murie decided that they fell into the Siberian class of dogs. Any slight differences between the Ipiutak dog skulls and the Siberian and Alaskan dog skulls with which they were compared were within the size variation considered to be normal for domesticated dogs of this type. One animal in particular (No. H43) was considered to be a dog-wolf hybrid, as the large size and heavier mandible and teeth approached those of a wolf rather than a dog. Murie concluded that two of the skulls represented the so-called Siberian husky type, three may have been variants of this same type, and the other animal (No. H43) was the hybrid mentioned above. He believed that there was adequate evidence that the Ipiutak dogs were of Asian origin.

Both sides of the Bering Strait have seen numerous explorations and excavations searching for the earliest evidence of habitation sites of the Asian migrants onto the North American continent. Unfortunately, very little relating to vertebrate remains has been reported. This may be due partly to poor preservational conditions which characterize this region and which have resulted in an extremely small sample upon which to base our interpretations. However, the subordinate role of organic remains in many early archaeological interpretations has also played a significant part in this gap in our knowledge of fossil vertebrates of all taxons from archaeological contexts. An example of this subordinate role can be seen in the 331-page report on the archaeology of Cape Denbigh (Giddings 1964). The text and 73 plates are devoted to a thorough coverage of nearly all aspects of the excavations on Cape Denbigh, but the recovered faunal remains are listed in a seven-line table in one paragraph. No scientific taxonomy is given for these animals, and some remains are classified simply as "bird" or "other." Obviously, it is quite possible to overlook many comparatively small canid fragments if one's interests are funneled in other directions. It is also possible that canids were not collected because it may have been assumed that they represented the common, local wolves, when in actuality they might have represented primitive dogs, as well as wolves or crosses between wolves and dogs.

An intriguing collection of seven large canid skulls that may represent both domestic dogs and wolves or even wolf-dog crosses were collected at the Bagnell site, which is a two-component village situated on the edge of a

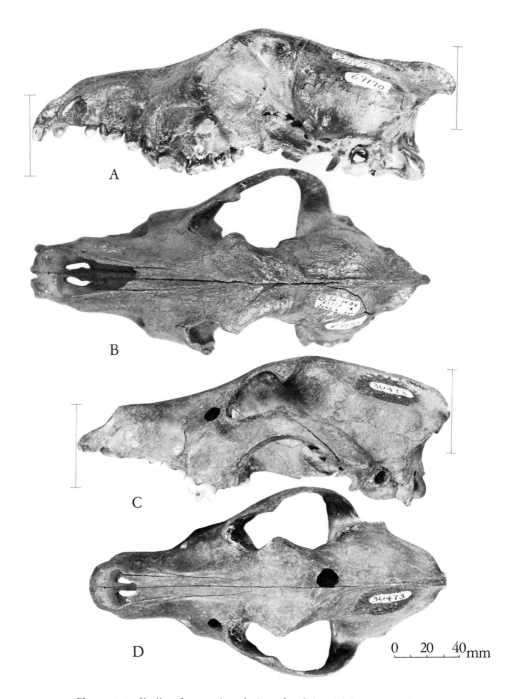

Figure 2.2. Skulls of normal and short-faced late Pleistocene wolves, *Canis lupus*, from Fairbanks, Alaska, area: A) normal adult skull, left lateral aspect; B) normal adult skull, dorsal aspect; C) short-faced wolf skull, left lateral aspect; D) short-faced wolf skull, dorsal aspect.

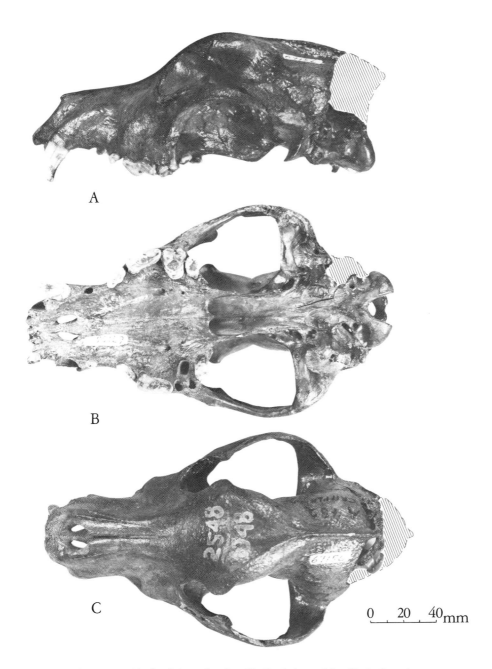

A

B

C

0   20   40 mm

Figure 2.3. Skull of short-faced wolf, *Canis lupus ?familiaris,* from late Pleistocene beds of Cripple Creek, Alaska: A) left lateral aspect of skull; B) palatal aspect of skull; C) dorsal aspect of skull. (Missing area indicated by cross-hatching.)

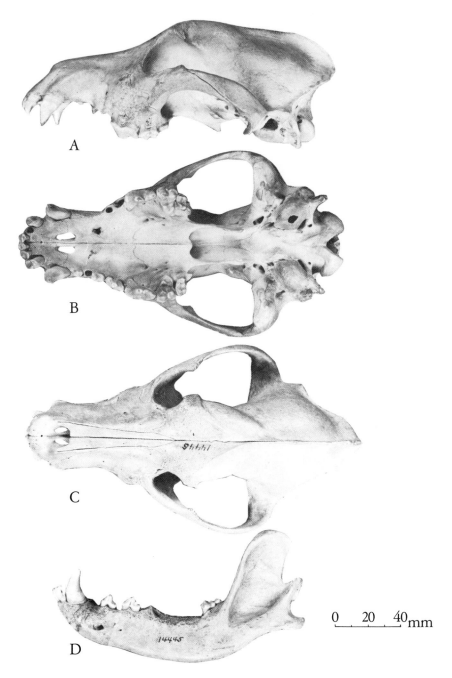

Figure 2.4. Skull and mandible of a Greenland Eskimo dog, *Canis famil-iaris*, collected by Peary Expedition in 1896: A) left lateral aspect of skull; B) palatal aspect of skull; C) dorsal aspect of skull; D) lateral aspect of left mandible.

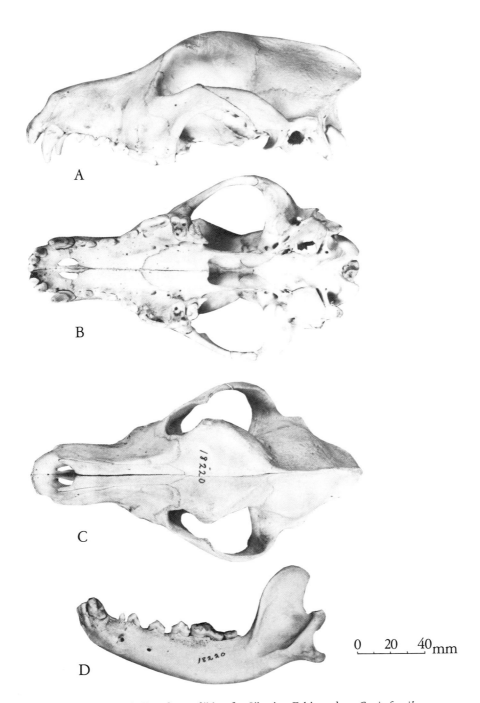

A

B

C

0 20 40 mm

D

Figure 2.5. Skull and mandible of a Siberian Eskimo dog, *Canis famil-iaris,* from northern Siberia, collected in 1901: A) left lateral aspect of skull; B) palatal aspect of skull; C) dorsal aspect of skull; D) lateral aspect of left mandible.

27

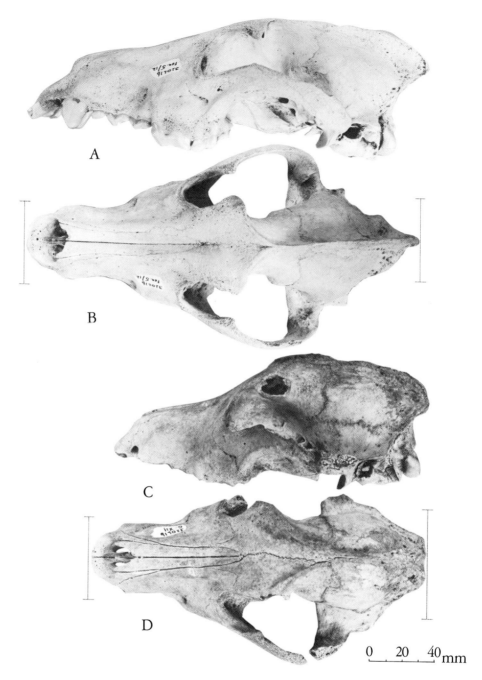

Figure 2.6. Skulls of normal and short-faced wolves, *Canis lupus,* from Bagnell archaeological site, North Dakota: A) left lateral aspect of normal skull; B) dorsal aspect of normal skull; C) left lateral aspect of short-faced skull; D) dorsal aspect of short-faced skull.

terrace above the Missouri River flood plain just north of Sanger, North Dakota (Fig. 2.6). The dates of the proveniences from which the skulls were collected are from as early as c. A.D. 1590, with the latest occupation beginning about c. A.D. 1700 (Lehmer 1973). Two of the specimens from the earliest level are domestic dogs of wolf size. Two are definitely wolf. From the later time of occupation, one skull is of a domestic dog, also of wolf size, and the remaining two are of wolves. A most intriguing observation is that one of the domestic dog skulls from the early layer possesses two strong wolf characteristics and seven morphological characteristics that are typical of large, domestic dogs. Admittedly there is a problem of temporal control for these proveniences; however, at this time it doesn't seem likely that the earliest proveniences contain any intrusive mixtures of dogs that were introduced by Europeans. Large, wolf-size dogs, particularly those which appear to be wolf-dog hybrids, are so rare in an archaeological context that the Bagnell canids are worthy of note.

Another published account of a large, hybrid cross of a wolf and an Indian dog is that contained in the report of excavations of a bison-kill site in southwestern New Mexico (Speth and Parry 1980). This find, a single canid skull, is illustrated and compared by Walker in the report by Speth and Parry (1980). Unfortunately, no provenience is given for this specimen. The overall range of dates for all proveniences on the site are from as early as A.D. 1420 ± 125 to A.D. 1845 ± 100. This canid would be of interest in relation to early taming of the wolf and, perhaps, its crossing with smaller Indian dogs only if it were established that it was collected from a provenience that predated the European introduction of large mastiffs or hounds. Since the range of dates is so great, it could easily be the result of a more recent crossing of a European dog and local wolf. The occurrence is recorded here as a plea to future workers to include all pertinent data and not just that which pertains to comparative zoology.

3. EARLY PREHISTORIC DOGS IN NORTH AMERICA

To date, the oldest substantiated finds of prehistoric domestic dogs in North America are those from Jaguar Cave, Idaho. Fossil fragments, consisting of incomplete mandibles, a single left mandible, and a small portion of a left maxilla, were all collected from excavations in an early Holocene rock-shelter in the Beaverhead Mountains of Lemhi County, Idaho (Figs. 3.1, 3.2). The excavations were conducted by a joint expedition of the Peabody Museum of Archaeology and Ethnology at Harvard University and the Idaho State Museum at the Idaho State University in 1961 and 1962. Carbon-14 dates indicate that the age of the deposits ranges from about 9500 to 8400 B.C. (Lawrence 1967). The site was determined to be a hunting camp (Sadek-Kooros 1972). An interesting aspect that remains unexplained concerning the canid from this site is its unwolflike appearance (Lawrence 1967), which suggests there may be an even earlier form, as yet undiscovered, that may link these Jaguar Cave dogs with a wild, ancestral form. One would not expect to find these early dogs in a locality so far south as the Jaguar Cave rock-shelter without finding similar remains in sites closer to the Bering Strait. Finding the remains described above was due to the discovery and excavation of a rock-shelter site; less-inviting sites, as yet unknown, between Jaguar Cave and the Bering Strait may, of course, hold equally important early dogs.

The morphological features that are present in the Jaguar Cave dog fragments are characteristically the same as those that are present in known specimens of *Canis familiaris*. For example, critical measurements taken of all fragments indicate that they are too small to be derived from the wolf. They were then compared to the coyote. They differ from this smaller, wild canid in being more massive, deeper dorsoventrally, and thicker lateromedially.

31

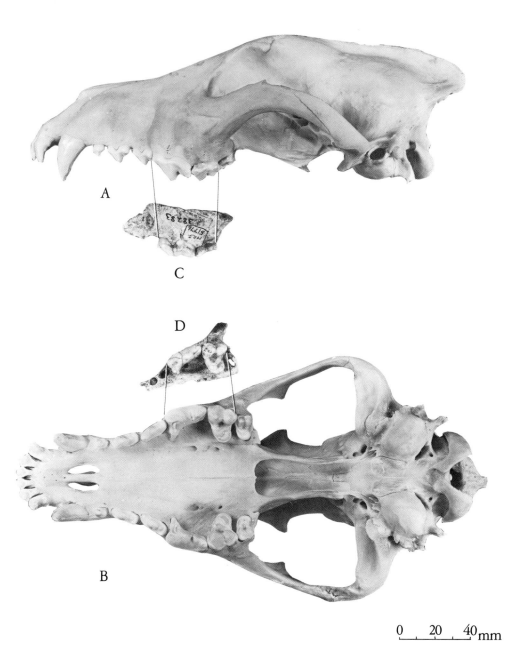

Figure 3.1. Skull of recent wolf, *Canis lupus lycaon* ♀, and maxillary fragment of Jaguar Cave domestic dog, *Canis familiaris:* A) left lateral aspect of wolf skull; B) palatal aspect of wolf skull; C) left lateral aspect of maxillary fragment of dog; D) palatal aspect of maxillary fragment of dog.

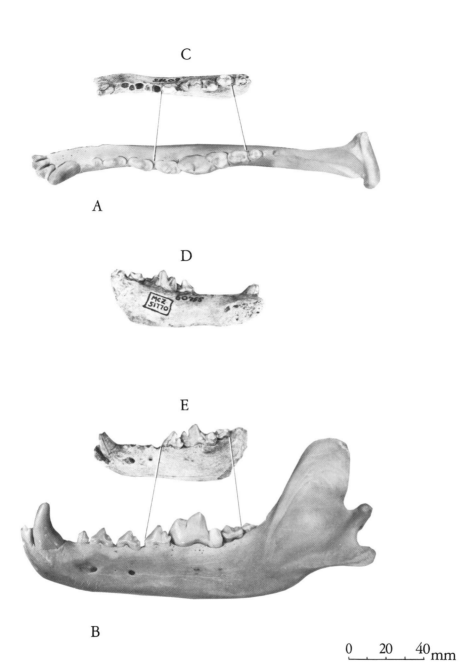

Figure 3.2. Mandible of a recent wolf, *Canis lupus lycaon*, ♀, and mandibular fragment of left mandible of a Jaguar Cave domestic dog, *Canis familiaris:* A) left mandible, occlusal aspect, of wolf; B) left mandible, lateral aspect, of wolf; C) mandibular fragment, occlusal aspect, of dog; D) left mandibular fragment, lingual aspect, of dog; E) left mandibular fragment, labial aspect, of dog. Specimens are from the same individuals as shown in Figure 3.1.

The tooth rows are short when compared to the size of the individual teeth, and this shortening of the jaws is accompanied by crowding of the tooth row. This aspect is particularly noticeable in the area of anterior premolars, where the alveoli do not lie in a straight line; they are, rather, set obliquely. The muzzle, as far as could be projected from the fragments, is more short-ened than in the coyote, and this characteristic seems to be well developed, as in later domestic dogs.

Some lower jaws of what are surely coyotes also occurred in the same site. These were determined to be of the local subspecies, *Canis latrans lestes*. Canid material from a considerably later site (also in Birch Creek Valley, Idaho), having a ^{14}C date not earlier than 2500 B.C., was also described by Lawrence (1968) as belonging to domestic dogs. The specimens included four nearly complete mandibles, a number of mandibular fragments, one broken skull, and two cranial fragments. The specimens from this later site were determined to be very similar in form and size to Eskimo dogs, with which they were compared in the Museum of Comparative Zoology at Harvard University. By inference, the specimens from Jaguar Cave were also from this group of dogs.

There exists a considerable temporal gap between the Jaguar Cave dogs and later prehistoric dogs from other areas in North America. Fossils from areas in the Southwest date from at least the time of Christ (Guernsey and Kidder 1921). From other areas they date perhaps from 8400 B.C., if we consider the canid burial from the Western Ozark Highland (McMillan 1970) as evidence. The dog from the Koster site in Illinois (Struever and Holton 1979) has an assigned date of 6500 B.C. Dogs from White's Mound in Rich-mond County, Georgia (Fig. 3.3), have been given a date of about 500 B.C. (Olsen 1970). Hill (1972) briefly noted a Middle Archaic dog burial from Illinois with an accompanying date of 5000 B.C.

G. M. Allen (1920) published the first comprehensive discussion of early dogs that were associated with the prehistoric peoples of the Western Hemisphere. This publication, unfortunately, is now out of date as well as out of print. Most of the critical finds relating to the development of the domestic dog were discovered subsequent to Allen's work. W. C. Haag (1948) based his evaluations of an osteometric analysis of some eastern prehistoric dogs on Allen's early publication. Haag's monograph is also out of date, as this one will be, too, as new discoveries are made in the future.

The monograph by Allen was for many years the standard work that was used to classify domestic dogs that were found in association with pre-historic human cultures. Allen listed these dogs as falling into the following groups, determined by comparative skull measurements and form (Allen 1920:

34

503): "(1) a large, broad-muzzled Eskimo dog, (2) a larger and (3) a smaller Indian dog, from which are probably to be derived several distinct local breeds. Of the larger style of dog, as many as eleven varieties may be distinguished; of the smaller, five." Some years ago I grouped southwestern prehistoric dogs into two groups, small and large, following, more or less, Allen's (1920) classification. The animals in the small group were fox-terrier–sized and were further refined or separated into small, short-faced, and small, long-faced dogs. The animals in the large group were long-faced and were comparable in size to the local coyote but a bit heavier in overall proportions (Fig. 3.4). This latter group was referred to by Allen as the large Pueblo Indian dog, or the Plains Indian dog.

After examining later published reports and newer finds and reexamining the older finds of Allen's time, it now appears that these groupings are, in a sense, artificial—particularly for dogs from the southwestern United States. The groups actually grade into one another in size, form, and amount of morphological variation, if a large enough collection is examined. The result is a more or less single mongrel group of southwestern Indian dogs. However, the entire range of size and form variation is not found in every archaeological site. It is still possible to find representatives of only a part of the spectrum—either small, short-faced or long-faced dogs, or large Pueblo Indians dogs—at specified sites. The overlap is at the extremes of each of these size groups and is quite logical, since the prehistoric Indian dogs were not registered American Kennel Club breeds, but were, instead, free-breeding, socializing mongrels. It is, of course, quite simple to pick out representative animals of these differing groups from collections of excavated canids if there are enough individuals assembled from a large number of archaeological sites. The Basketmaker and early Pueblo Indian dogs of the Southwest and those from White's Mound, Georgia, as well as the dogs from the shell-heaps of Kentucky or Alabama all show a close similarity in size and form. All of these animals are quite advanced domestic dogs, and most are of a comparatively small size, although—as stated earlier—there are exceptions to this rule (for example, both small and large forms of dogs are found in the same stratigraphic level at the Jaguar Cave site).

A singular find in North America, and fortunately one of the oldest (around the time of Christ), was the discovery by Guernsey and Kidder (1921) of two well-preserved mummies of Basketmaker dogs (Fig. 3.5). Both are from White Dog Cave in the northwestern part of Arizona near the town of Kayenta. One dog is a long-haired animal (actually a buff color, not white) about the size of a small collie, with erect ears and a bushy tail. The other dog is considerably smaller, black and white, about the size of a terrier, with a

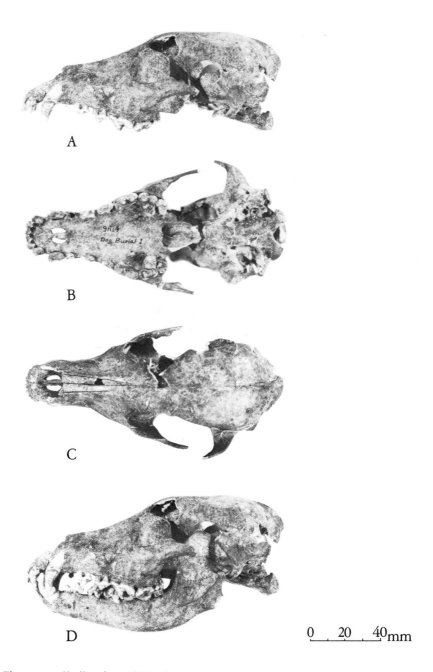

0 20 40mm

Figure 3.3. Skull and mandible of a Neolithic domestic dog, *Canis famil-*
iaris, from White's Mound archaeological site, Georgia:
A) left lateral aspect of skull; B) palatal aspect of skull; C)
dorsal aspect of skull; D) left lateral aspect of articulated
skull and mandible.

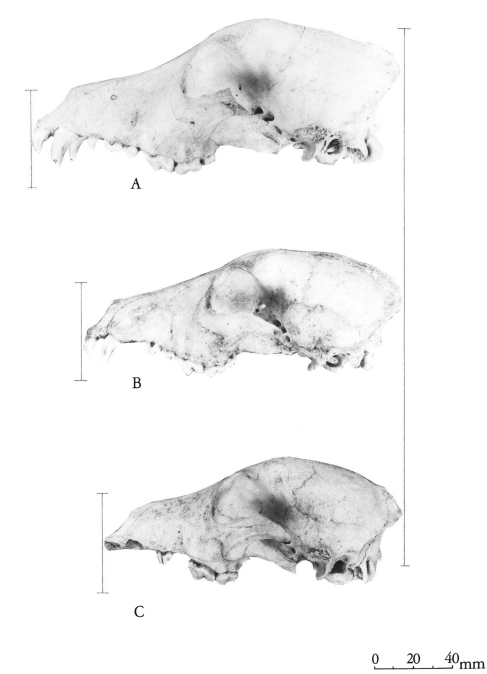

0   20   40mm

Figure 3.4. Skulls of three common types of prehistoric southwestern Pueblo Indian dogs, *Canis familiaris:* A) large, long-faced Pueblo Indian dog (Mesa Verde Ruins, Colorado), left lateral aspect of skull; B) small, long-faced Pueblo Indian dog (Pecos Pueblo, New Mexico), left lateral aspect of skull; C) small, short faced Pueblo Indian dog (Zuni Pueblo, New Mexico), left lateral aspect of skull.

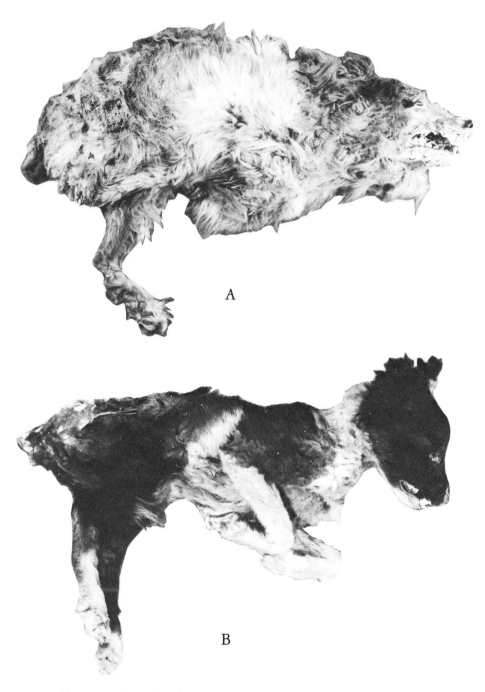

A

B

Figure 3.5. Mummies of early Basketmaker Indian domestic dogs, *Canis familiaris,* from White Dog Cave, Marsh Pass, Arizona: A) white dog; B) black and white dog.

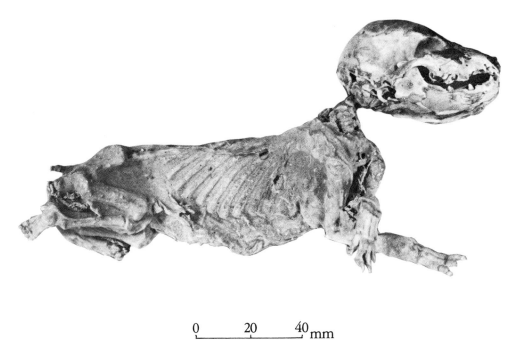

0   20   40 mm

Figure 3.6. Mummy of domestic dog (puppy), *Canis familiaris*, from Antelope House Ruin archaeological site, Canyon de Chelly, Arizona. Associated with prehistoric culture from Basketmaker to Pueblo I.

short-haired but shaggy coat; the ears are erect, the tail long and bushy, and the muzzle is noticeably shortened. Both dogs were found in association with a number of fur-robe-wrapped human mummies, one of which had a finely woven tumpline and pad made of dog hair within its wrappings.

Many early and later Pueblo Indian sites have yielded dog remains. In particular, reference is made to immature puppies (Fig. 3.6), as well as adults, at Antelope House in Canyon de Chelly, Arizona (Kelley 1975), found in levels dating from about A.D. 1125 to 1175. At the 14th century Grasshopper Pueblo in east-central Arizona, a number of small domestic dogs were recovered during sixteen years of excavation by the University of Arizona Archaeological Field School (Olsen 1980). A unique find of an immature gray wolf, *Canis lupus*, consisting of a right premaxilla and deciduous dentition, was also recovered from one of the rooms of this pueblo. The implication suggested by this fragment of a wolf pup is that it may have been kept as a pet, perhaps with a view toward taming. This interesting idea cannot, of course, be substantiated beyond its being a possibility.

Nearly every Pueblo excavation in the southwestern United States has produced some evidence of the domestication of the dog. These sites date from A.D. 1050 to A.D. 1273, and range in size from Cliff Palace Ruin in Mesa Verde National Park, Colorado, with its multi-storied masonry construction of many rooms and great stone towers, to the more modest pueblos of fewer rooms at Keetseel and Betatakin in the area of Monument Valley, Arizona. Tree-ring dates for Keetseel range between A.D. 1274 and A.D. 1284; those for Betatakin are between A.D. 1260 and A.D. 1277. By the time of these late dates, the domestic dogs were well advanced and were morphologically the same as present-day dogs.

As of 1985 we have not recovered the ancestral forms of *Canis familiaris* that would help to bridge the gap between small, late Paleolithic wolves and early Neolithic dogs from Asia and from Jaguar Cave, Idaho, or the gap from the Jaguar Cave dogs to the well-known series of dogs of the southwestern United States that date from the time of Christ and continue to the present time.

4. THE SMALL PLEISTOCENE WOLVES OF CHINA AS ANCESTRAL DOGS

It is generally accepted among workers concerned with the problem of the beginnings of canid domestication that the basic ancestral animal was a small subspecies of the wolf, *Canis lupus* (Olsen and Olsen 1977; Lawrence 1967), and that China is one of the most likely areas of development. Tracing the biological relationship among the living species of canids is less difficult than working only with paleontological or archaeological remains, as one has to do in relation to Paleolithic or Neolithic sites. Data that are present in living animals, such as genetic relationships, voice patterns, gait, and location of sweat glands, can be compared and evaluated for possible generic associations. On the other hand, with Paleolithic or Neolithic animals, even the best-known materials—fragmentary skulls and dentitions—are rare. Complete skeletons are virtually unknown, and fragmentary post-cranial elements, when found, are of limited taxonomic value because they generally cannot be reliably determined to a specific level of classification. Among the specimens of *Canis lupus* associated with Pleistocene hominid remains in the Old World, several are known from China. There are enough individuals available for valid and meaningful comparisons of the preserved morphological characters, so that a multivariate analysis can be undertaken to also include other related species of *Canis*, both fossil and living.

Neolithic domestic dogs from both China and North America are rather numerous and are represented by fairly complete skeletal material, which allows for detailed comparisons. A major problem arises, however, when one tries to bridge the gap between these small, Neolithic dogs and the early, large, wild wolves of North America. There is a decided size differential

41

between the earliest known Neolithic dogs and the earlier wolves. An early, smaller wolf must be sought (Olsen and Olsen 1977) as a possible, more closely allied progenitor.

There are 32 subspecies of the present-day wolf, *Canis lupus*, recorded world-wide (Mech 1974). Criteria used for separating and establishing these many subspecies are based mainly on body size, form, color of pelage, and geographical distribution. Whether these many listed races or subspecies constitute valid taxonomic distinctions is certainly open to question and discussion. The North American subspecies, particularly the arctic forms, are among the largest of the wolves, showing little change in size or form since Pleistocene times, while the Chinese wolf and some Asian subspecies are among the smallest (Fig. 4.1). Modern taxonomic works published in North America often indicate discrepancies and inconsistencies regarding the correct name of this wolf from China. It was given the name *Canis chanco* by Gray (1863). Mivart (1890), in his monograph on the Canidae, was the first worker to give the Chinese wolf the subspecific designation of *chanco*, although he listed it as "*Canis lupus* variety *chanco*." Allen (1938), in his extensive coverage of Chinese and Mongolian mammals, reclassified it again to the subspecific level of *Canis lupus chanco*. Mech (1970) listed only *Canis lupus laniger* as being valid for the Chinese animal. However, in 1974 he changed his classification to *Canis lupus chanco*. I accept the classification *Canis lupus chanco* as being valid, based principally on its being accredited by taxonomists at the Institute of Zoology, Academia Sinica, in Beijing, the People's Republic of China (Wang Sung, pers. comm., August 1980), and also on Allen's and Mech's classifications.

One of the earliest known associations of *Canis lupus* with hominids (*Homo erectus pekinensis*) is from the fossil site at Zhoukoudian, located about 42 kilometers southwest of Beijing in the People's Republic of China. The associated remains are approximately 500,000 to 200,000 years old. Although this association of hominids and wolves at this early period does not imply in any way either taming or early domestication, it does place both genera of animals in contemporary association that apparently continued until such time that these events did occur.

A number of small wolves have been described from the Pleistocene deposits of China. Zdansky (1924) assigned the name *Canis chihliensis* to a small, extinct wolf. However, Allen (1938) believed that this animal was within the accepted range of variation of the present-day wolf, *Canis lupus chanco*, and placed it in synonymy with that subspecies. Osborn (1931) published a misleading statement in regard to canids occurring at Zhoukoudian, Locality 1. He stated that the "Fossil dog (*Canis sinensis*)" was present in the

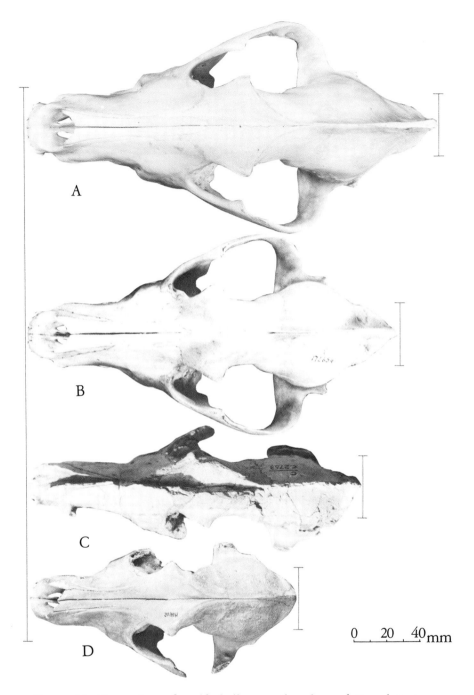

Figure 4.1. Comparison of canid skull proportions in modern wolves, a Pleistocene wolf of China, and a Neolithic domestic dog of China: A) modern Western-Hemisphere wolf, *Canis lupus lycaon,* dorsal aspect of skull; B) modern Chinese wolf, *Canis lupus chanco,* dorsal aspect of skull; C) late Pleistocene Chinese wolf, *Canis lupus variabilis,* from archaeological site of Zhoukoudian, dorsal aspect of skull; D) Neolithic Chinese domestic dog, *Canis familiaris,* from archaeological site of Hemudu, dorsal aspect of skull.

lowest layers of the cave deposits. Pei (1934) discussed this small canid under the name *Canis (Nyctereutes) sinensis*. Pei also credited Teilhard and Piveteau (1930) with making this name change from *?Vulpes sinensis*, as had been listed earlier by Zdansky (1927). Zdansky gave his reasons for changing the taxonomy as disagreeing with Schlosser's (1903) assignment of this same animal to *Vulpes sinensis*.

Allen (1938), in his revision of the canids of China and Mongolia, stated that the genus *Nyctereutes* included only the small, fox-like raccoon-dogs. He went on to say (p. 348) that they were more nearly allied to the Arctic fox, *Alopex*. He credited Brass (1904) with assigning the name *Nyctereutes sinensis* to this animal, but said that more properly it was synonymous with the living raccoon-dog *Nyctereutes procyonoides*. Rather than to indulge in taxonomic semantics, based on incomplete or inadequate material, I prefer to state that these small Zhoukoudian canids are, in all probability, ancestral to the single, living species of raccoon- dog, *Nyctereutes procyonoides*, rather than being in the direct line of canid evolution leading to the domestic dog, *Canis familiaris*. A reexamination of a number of Pleistocene canid skulls in the collections of the Institute of Vertebrate Paleontology and Paleo-anthropology in Beijing, the museum at Zhoukoudian, and the Tianjin Natural History Museum has indicated that a more thorough and lengthy study of all the small Pleistocene canids should be undertaken. The conclusions of such a study might possibly change or alter some of the taxonomic assignments that are now associated with some specimens and would result in the synonymy of some that are now carried under separate taxonomic listings.

Pei (1934) established a new subspecies, *Canis lupus variabilis* (although he listed it as a "variety") for the Zhoukoudian small wolves (see Figs. 1.5 and 4.16) that were found in association with lithic evidence attributed to *Homo erectus pekinensis*. I have carefully measured wolf skulls of this small subspecies from Localities 1, 3, and 13 at Zhoukoudian and have compared them with other critical canid material in an effort to determine the position of this animal in the evolutionary progression to the recent canids. *Canis lupus variabilis* maintains its heavy, wolf-like proportions, even though it is a considerably smaller animal overall from any living subspecies of *Canis lupus*. The sagittal crest, usually quite prominent in all subspecies of living wolves, is noticeably reduced in the known specimens of *Canis lupus variabilis*. The diagnostic characteristics separating this wolf from others were stated by Pei (1934) as "size moderate M_1 varies from 22–24 mm in length." I agree with Pei's careful assessment of the morphological values of these small wolves that were made after he compared them with the living Chinese wolf, *Canis lupus chanco*. Pei did not have the advantage of comparing the specimens collected subsequent to his early study, particularly those from Locality 13 (Pei 1936; Teilhard and Pei 1941). However, he correctly stated (1934:17): "Although no sharp line can be traced between the above described *Canis* and a true *lupus*, the marked differences found in the size, and in the cranial characters, seem to be sufficient for creating, at least, a new variety, *Canis lupus variabilis*, for the Zhoukoudian Locality 1 small wolf." Pei (1934:18) also believed that the Nihewan wolves attributed to *Canis chihliensis* by Teilhard and Pivetean (1930) should also more properly be included under his new category of *Canis lupus variabilis*. If so, we then have a problem of taxonomic priority regarding which of the two specific names should be considered valid under the Rules of International Zoological Nomenclature. For the present, it is less confusing to follow Pei (1934) and to refer to these small Zhoukoudian wolves as *Canis lupus variabilis*. *Canis lupus variabilis*, as reported by Kahlke and Chow (1961), has also been found in layers 1 through 10 at Locality 1, Zhoukoudian, associated with *Homo erectus pekinensis*. The animal was also reported from layer 11, from which no hominid remains were recovered. The same small wolf is also known from Lantian in Shaanxi Province, People's Republic of China, where it is associated with a fauna that is considered slightly older than the Zhoukoudian fauna (Hu and Qi 1978).

The wolf was already a part of the natural faunal assemblage associated with early Chinese hominids, which adds to the likelihood of its being a candidate for continued association through the Paleolithic and into the Neolithic, by which time it was certainly domesticated.

5. Prehistoric Dogs in Mainland East Asia
John W. Olsen

The topographically complex region of mainland east Asia has undergone extreme changes in the past ten to fifteen thousand years with regard to the quantity and distribution of potentially habitable land, due to the fluctuation of sea levels associated with the climatic perturbations of the Pleistocene. The problem of viewing the eastern half of the Old World as a rather homogeneous zone in terms of its prehistoric sequences has been compounded for Western scholars by the paucity of English-language reference materials dealing with the archaeology of this immense territory. Research conducted from the 1950s to the 1980s, however, has demonstrated unequivocally that many regions of East Asia are characterized by extremely diverse and unique archaeological assemblages. In particular, several centers of early food production, considered to be autochthonous developments, are known in East Asia (Chang 1977; Bellwood 1978; Solheim 1972; and others). These centers emphasize the need to reexamine the eastern Old World for potentially important evidence of early animal husbandry.

The background for the origins of canid domestication was discussed in Chapters 1 through 4, which examined the most plausible path of phylogenetic development that has culminated in the domestic dog. Based upon the contention that the wolf (*Canis lupus*) is the most likely progenitor of the modern domestic dog, it is not surprising that the East Asian landmass and its insular neighbors should produce evidence of canid domestication in early Neolithic contexts. As noted earlier, several wild species of *Canis lupus* are known to have occurred in East Asia during the late Pleistocene, when it is assumed that human hunters first began to view their canid counterparts in terms other than those of competitor and enemy. Unfortunately, the late Pleistocene archaeological record has not yet produced unquestionable

comparison led Zhou (1981:341–393) to conclude that these remains recovered at Cishan may be positively regarded as domestic. Since the Neolithic component at Cishan is thought to predate the earliest previously known agricultural complex in north China—the Yangshao Culture (An 1979a, 1979b; Xia 1977:387), these canid remains constitute the oldest unequivocal traces of the domestic dog on mainland East Asia.

In Henan Province a pre-Yangshao Neolithic assemblage has been identified at Peiligang (Xinzheng County) which has a range of associated very early ^{14}C dates: 6435 ± 200, 7145 ± 300, 7185 ± 200, 7445 ± 200, 7885 ± 480, and 9300 ± 1000 B.P. (Laboratory of the Institute of Archaeology, CASS 1979: 90; 1980:373; 1981:365). The ceramic and lithic assemblages from Peiligang have been discussed in the Chinese literature (Kaifeng Prefecture Cultural Relics Management Team and Xinzheng County Cultural Relics Management Team 1978; Kaifeng Prefecture Cultural Relics Management Team et al. 1979; An 1979a); however, the analysis of faunal remains from this site was still in progress in the middle 1980s. Zhou (1980) reported that a domestic dog left mandible (specimen number T110-0-8, Fig. 5.1E) from Peiligang has been identified which possibly predates remains from Cishan, Hebei. However, we must await the comparison of this and other specimens with those from other Neolithic contexts in China before definitive conclusions may be reached.

One of the most important early Neolithic sites discovered during the 1970s in China is the village complex of Hemudu, located on the Luojiang Commune in Yüyao County, Zhejiang Province. Excavations, which commenced at Hemudu in 1973, revealed evidence of four stratigraphic units. The two lowermost strata bear archaeological materials associated with a previously unknown early Neolithic complex, known tentatively as the Hemudu Culture (Zhejiang Provincial Cultural Relics Management Team and Zhejiang Provincial Museum 1976, 1978). Radiocarbon determinations of organic material associated with the Hemudu Culture levels have yielded dates of 6310 ± 100 and 6065 ± 120 B.P. (Xia 1977). These determinations suggest a much earlier date for the origins of rice cultivation on China's Pacific seaboard than was previously suspected.

Specialized examinations of the plant and animal remains from Hemudu have been published in a summary report (Zhejiang Provincial Museum, Natural History Section 1978). In addition, scholars at Zhejiang Provincial Museum in Hangzhou lent the canid crania from Hemudu for comparative research at the University of Arizona (Olsen, Olsen and Qi 1980). The remains of six dogs have been uncovered at Hemudu, all of which are rather small animals with foreshortened muzzles and correspondingly crowded dentitions (Fig. 5.2A–C).

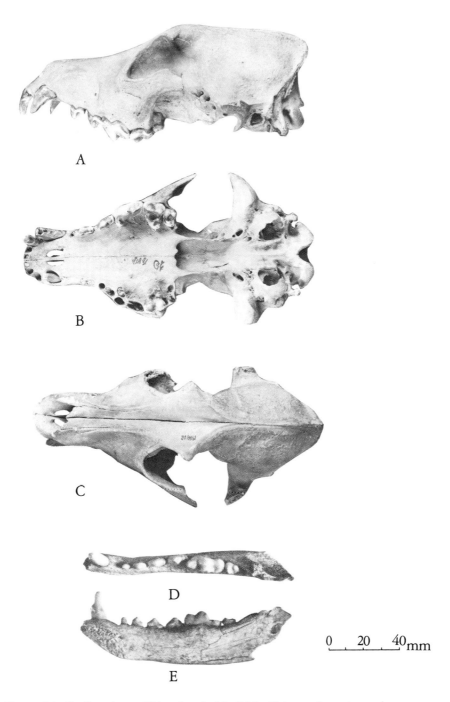

A

B

C

D

E

0 20 40 mm

Figure 5.2. Skull and mandible of early Neolithic Chinese dogs. An early Neolithic domestic dog, *Canis familiaris,* from Hemudu, Zhejiang: A) left lateral aspect of skull; B) palatal aspect of skull; C) dorsal aspect of skull. An early Neolithic domestic dog, *Canis familiaris,* from Banpo, Shaanxi: D) occlusal aspect of right mandible; E) medial aspect of right mandible.

51

The well-developed agricultural communities of the Yangshao Culture which characterized the north Chinese loess-lands from roughly 5000 to 3800 B.C. provide the earliest relatively large sample of domestic canid remains from a rather restricted cultural and temporal context. Scores of Yangshao Culture sites have been excavated, and all but a few have yielded evidence of extensive reliance upon domestic dogs and pigs (*Sus scrofa*) in the agricultural subsistence base.

In the mid-1980s the best-known early Yangshao site was that of Banpo, located in the eastern suburbs of Xi'an, Shaanxi Province in the Wei River valley (Institute of Archaeology et al. 1963). Banpo, a large community comprising about 50,000 square meters (of which about 20 percent has been excavated) has yielded a spectrum of cultural and ecological remains of great significance in the analysis of early agriculture in China. Radiocarbon dates and stratigraphic evidence suggest that several periods of occupation account for the accumulation of cultural remains at Banpo. Analysis of excavated material occurring in various stratigraphic levels indicates the employment of swidden agriculture. The radiocarbon dates derived from the site suggest a period of occupation of from roughly 4200 to 3600 B.C. (Chang 1977:84, 485; Xia 1977).

Li and Han (1959; reprinted 1963) have published extensively on the zooarchaeological remains uncovered at Banpo, including a consideration of the domesticates which are present. At least five individual domestic dogs are represented in the study of Li and Han (1959), although additional criteria, such as size, suggest that several more individuals may be present.

Having made a morphometric comparison of the Banpo canids with *Canis lupus* osteological material, Li and Han (1959:176) concluded: "Based upon the relatively small skull, protruding rostrum, small carnassials, and the curved inferior margin of the mandible, obviously distinguishing it from the wolf (*Canis lupus*), the specimen is undoubtedly that of a dog." The following mandibular measurements for the *Canis familiaris* specimen number 276.2 from Banpo are provided by Li and Han:

C to $M_{\overline{2}}$	Length 79.5 mm
$P_{\overline{1}}$ to $M_{\overline{2}}$	Length 62.3 mm
$M_{\overline{1}}$	Length 20.5 mm; width 7.9 mm

The question of the use of dogs at Banpo and other early Neolithic sites in China is as yet unresolved; however, the recovery of *Canis familiaris* bones from many depositional contexts in these sites, including refuse middens, suggests that they may have been used as a source of protein, in addition

52

to non-subsistence functions. The extent to which Chinese Neolithic subsistence was dependent upon the dog and the pig was uncertain in the mid-1980s, because a quantitative study of physical evidence for the processing of dogs for consumption (butchery marks, patterned burning, differential representation of elements) had not yet been conducted.

Domestic dogs similar to those described from Banpo have, of course, been reported from dozens of other Yangshao Culture sites, including Jiangzhai, Shaanxi, Jingcun, Shanxi, Miaodigou I, Henan; Sanliqiao, Henan, and Zhongzhoulu, near Luoyang, Henan (Ho 1975:93). Unfortunately, no detailed analytical studies had been performed on these dog remains as of 1984, so we can only note that these specimens appear to closely resemble the small dogs from Banpo (see Fig. 5.2D, E) in terms of size and gross morphology. The other early agricultural complexes of eastern China —the Qingliangang of the Yangzi Delta region and the Dapenkeng of the southeastern coast and Taiwan province—also exhibit a reliance on pigs and dogs in their agricultural repertoire. Dog remains are reported from nearly every Qingliangang and Dapenkeng site that has been excavated; unfortunately, we have little more than mere statements of their presence to base interpretations upon.

After roughly 3200 B.C. the late Neolithic cultures of China, referred to collectively as Longshanoid (Chang 1977:144, 154–155), begin to exhibit features which are broadly similar across China. An interesting aspect of these later Neolithic cultures is the rather sharp increase in the range and quantity of domesticates present in most sites. Cattle and sheep take their places alongside pigs and dogs as economically important animals. Nearly all of the Longshanoid sites excavated through 1984 have produced remains of domestic dogs (Zhou 1980); consequently, only a few will be mentioned in detail here.

The site of Xiawanggang in Xichuan County, southwestern Henan Province, has produced dog remains in a rather good state of preservation that are clearly smaller than those of the local Chinese wolf (Jia and Zhang 1977:44). Dog remains from the middle and late Qüjialing (Longshanoid) Culture strata at this site were examined (Olsen, Olsen, and Qi 1980) and found to be significantly smaller than the living species of local wolf (Figs. 5.3, 5.4, 5.5, 5.6, 5.7D–H). This result supported the findings of the original investigators (Jia and Zhang 1977), who concluded that the Xiawanggang dogs from these late Neolithic contexts were so advanced that they must represent an animal far removed from the initial process of domestication. Even the dog remains recovered in association with a much earlier Yangshao Culture human burial at Xiawanggang (Henan Provincial Museum et al. 1972:7) appear indistinguishable from the Qüjialing dogs examined by Olsen and Olsen.

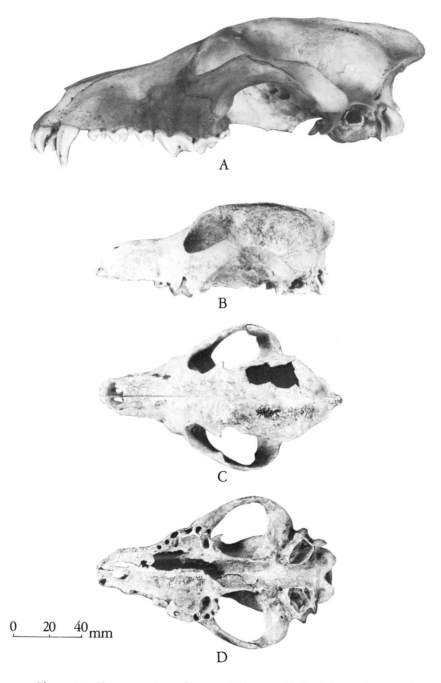

0 20 40 mm

Figure 5.3. Size comparison of a recent Chinese wolf, *Canis lupus chanco,* and a late Neolithic domestic dog, *Canis familiaris,* from the site at Xiawanggang, Henan: A) left lateral aspect of wolf skull; B) left lateral aspect of dog skull; C) dorsal aspect of dog skull; D) palatal aspect of dog skull.

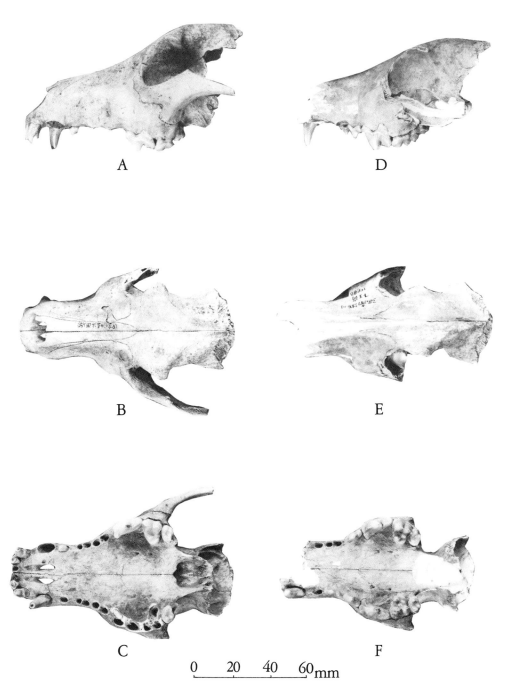

A   D

B   E

C   F

0   20   40   60 mm

Figure 5.4. Partial skulls of large (*left*) and small (*right*) Neolithic domestic dogs, *Canis familiaris* from the Longshanoid site of Xiawanggang, Henan. Large dog: A) left lateral aspect of anterior portion of skull; B) dorsal aspect of anterior portion of skull; C) palatal aspect of anterior portion of skull. Small dog: D) left lateral aspect of anterior portion of skull; E) dorsal aspect of anterior portion of skull; F) palatal aspect of anterior portion of skull.

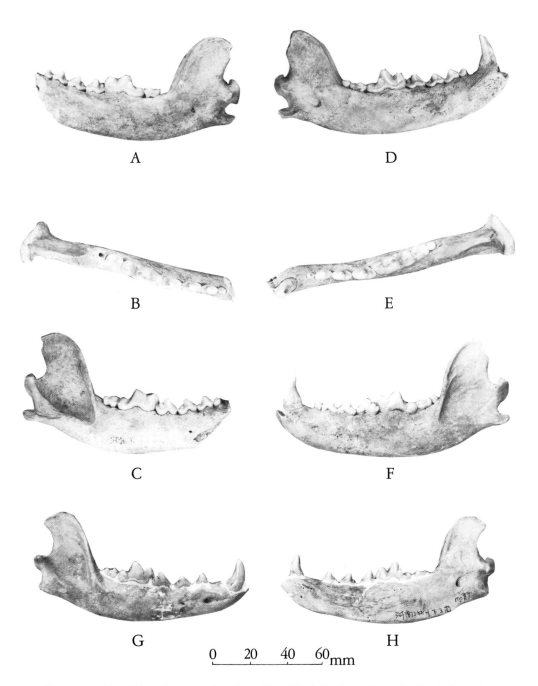

0 20 40 60mm

Figure 5.5. Mandibles of one small and two large Neolithic dogs, *Canis familiaris,* from the Longshanoid site of Xiawanggang, Henan. Right mandible of a large dog: A) medial aspect; B) occlusal aspect; C) lateral aspect. Left mandible of a large dog: D) medial aspect; E) occlusal aspect; F) lateral aspect. Right mandible of small dog: G) lateral aspect; H) medial aspect.

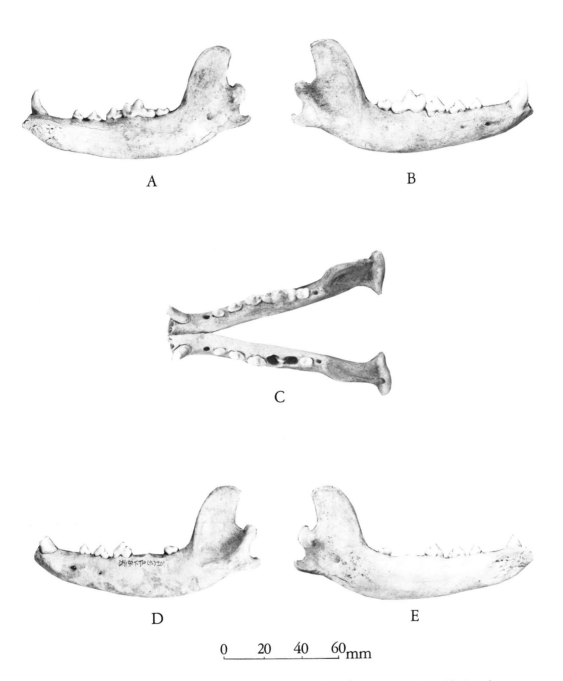

A

B

C

D

E

0   20   40   60 mm

Figure 5.6. Mandibles of a Neolithic dog, *Canis familiaris,* from the Longshanoid site of Xiawanggang, Henan: A) right mandible, medial aspect; B) right mandible, lateral aspect; C) right and left mandibles, occlusal aspect; D) left mandible, lateral aspect; E) left mandible, medial aspect.

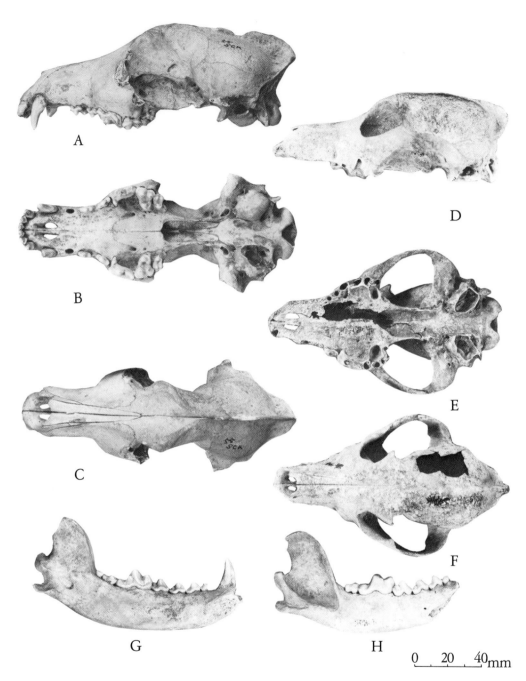

Figure 5.7. Skulls and mandibles of Neolithic Chinese dogs. Large domestic dog, *Canis familiaris,* from the Neolithic site of Kexingzhuang II, Shaanxi: A) left lateral aspect of skull; B) palatal aspect of skull; C) dorsal aspect of skull. Small domestic dog, *Canis familiaris,* from the Neolithic site of Xiawanggang, Henan: D) left lateral aspect of skull; E) palatal aspect of skull; F) dorsal aspect of skull. Mandibles of two large Neolithic domestic dogs, *Canis familiaris,* from Xiawanggang, Henan: G) medial aspect of a left mandible; H) lateral aspect of a right mandible.

Xiawanggang has produced a potentially vital sequence of domestic dog remains spanning the period from the early Yangshao Culture, perhaps 6200 years ago, to the Western Zhou Dynasty which ended in the eighth century B.C. Unfortunately, as of 1985 a detailed morphometric comparison of these dogs had not yet been undertaken.

The southeastern Chinese coast has also yielded remains of several local variants of the Longshanoid tradition (Chang 1977:168). At Tanshishan, in Minhou County, Fujian Province, an assemblage of late Neolithic remains, ^{14}C dated to slightly earlier than 3000 B.P., has been identified (Laboratory of the Institute of Archaeology, CASS 1974:337). Collected during excavations conducted in 1964 and 1965, the animal remains have been the subject of a short report (Qi 1977) which includes a discussion of the Tanshishan domestic dog remains in relation to those derived from other archaeological contexts. In her analysis of the Tanshishan fauna, Qi (1977:302) compares the mandibular morphometrics of one well-preserved specimen (see Fig. 5.1 A–D) with those of dogs from the Early Yangshao Banpo site and the late Shang Dynasty urban complex of Yin, near modern Anyang, Henan. On nearly all points, the Tanshishan canid appears smaller than its counterparts at both the Yangshao and Bronze Age sites.

The importance of this particular fossil find is two-fold. First, like other late Neolithic dogs in China, the specimen from Tanshishan was obviously derived from a fully domesticated individual that can in no way be considered a transitional form. Second, as Qi's (1977) analysis points out, there is an apparent variation in size among the domestic dogs present in early Neolithic through late Bronze-Age archaeological contexts in China; all of these dogs, however, are readily distinguishable from the local *Canis lupus* population by virtue of their smaller comparative size and other morphological features. The temporal and geographic factors which may have influenced this apparent size variation among domestic canids in prehistoric and early historic contexts in China constitute an important research area for the future.

Longshanoid sites all over China have produced remains of domestic dogs generally dating to the third millennium B.C. Kexingzhuang II in Shaanxi (see Fig. 5.7 A–C), Miaodigou II and Sanliqiao in Henan, Chengziyai in Shandong, Qüjialing in Hubei, and the sites of Liulin and Dadunzi in Jiangsu are typical examples of such late Neolithic sites which have yielded remains of domestic dogs. In the mid-1970s Ho (1975: 97) reported that no systematic description or interpretation of these canid remains had been undertaken; unfortunately, this situation remained the same in 1985. Hence, little can be stated with confidence beyond the mere presence of domestic dogs in these contexts.

In the Inner Mongolia Autonomous Region, the multicomponent Yangshao-Longshan site of Ashan, near Baotou on the Huanghe, has produced large quantities of apparently domestic canid remains (Fig. 5.8C–D) which were undergoing analysis at the Inner Mongolia Museum in Huhehot in the mid-1980s. The late prehistoric site of Dadianzi, in Aohan Banner, Inner Mongolia, has also produced *C. familiaris* remains, including a complete skull and mandibles (Fig. 5.8A–B) from a cultural stratum ^{14}C dated to 3685 ± 135 B.P. (Zhou 1980). Archaeological investigations in the 1970s and 1980s in Inner Mongolia have brought to light considerable quantities of domestic dog remains, a few of which are shown in Figure 5.9. The recovery of these remains, as well as those of other domestic animals—such as cattle, horses, pigs, sheep, and goats—is of interest, given the development of a pastoral, nomadic way of life which later came to dominate the steppes of central Asia by early historic times.

Obviously, the role played by domestic canids in a society oriented toward extensive herding may be quite different from that of dogs tied to communities engaged in intensive agriculture. Given the paucity of direct information currently available on the functions of domestic dogs in prehistoric societies in China, detailed morphometric comparison of canids from sites in frontier areas, such as Inner Mongolia and Xinjiang, with those derived from the more settled agricultural villages of the Huanghe valley, for example, may prove of interest in an attempt to explain apparent size differences in populations of prehistoric domestic dogs within China.

China's historic period—the origin of which current archaeological evidence indicates is coincident with the emergence of the first urbanized, bronze-using civilization of East Asia, the Shang—offers many examples of the continued importance of dogs in Chinese civilization. However, as this book deals with the origins of the domestic dog, rather than subsequent development, a single example drawn from the late Bronze Age will suffice as an indication of the degree to which dogs figured in the early historic Chinese civilization which emerged in its fully fledged form in the early second millennium B.C.

The archaeological remains of the urban complex known by the Shang as Yin, the last capital city of that dynasty, is located in the vicinity of Anyang in northern Henan Province. The city, which was inhabited from circa 1384 B.C. until the Zhou Conquest about 1100 B.C., has been the subject of intensive archaeological investigations since 1928. Between May 1969 and May 1977, 939 Shang tombs were excavated in the western portion of the ruins of the city of Yin (Anyang Archaeological Team, IA, CASS 1979). Most of the tombs were of modest size, only five being monumental enough to be provided with a tomb passage entrance. A number of tombs contained sacrificial victims (Fig. 5.10). Seventeen of the tombs yielded a total of 38 human sacrificial victims; other tombs contained the remains of sacrificed animals such as horses, pigs, goats, fish, and dogs. Of these, the most common was *Canis familiaris*: there were 439 sacrificial dogs in 339 of the excavated tombs (Anyang Archaeological Team, IA, CASS 1979:46–48). Only one dog burial was collected and saved (personal observation at Anyang, August 1984).

One hundred five of the excavated tombs produced dogs which were recovered from undifferentiated pit fill and 197 tombs (such as M9, M294, M326, and M703) were found to contain dogs carefully placed in the *yaokeng*, or "waist pit," a trench located below the waist of the human interment that was a specially prepared locus for the disposition of sacrificial animals (Fig. 5.10B, C). Dog crania were also found included in the researcher's inventory of burial goods placed on the *ercengtai* or "second-level platform" of the tomb construction (Fig. 5.10A), and in some cases (Tomb M613, for example) dog remains were found in the general pit fill, the *yaokeng,* and the *ercengtai.* These cases attest to the importance of the dog as a sacrificial animal in late Shang Yin. Many of the dog skeletons bore bronze bells suspended from their necks, and in one case (Tomb M785) the sacrificial dog was accompanied by both a bronze bell and jade plectra (Fig. 5.10B); these articles also indicate the importance placed on dogs as sacrificial victims, given the predominantly ceremonial use of bronze and jade artifacts during the Shang period. Some of these dog burials (Fig. 5.11) were examined by Stanley J. Olsen at the Institute of Archaeology in Beijing during the summer of 1980; all resembled small, well-advanced domestic forms, as reported by Zhou (1980).

Many other examples of the importance of the dog in Bronze-Age China are found in the archaeological record and in contemporary historical documents (Anyang Archaeological Team, IA, CASS 1979:48; Keightley 1978:85–87; Chang 1977:357). Such reports summarize the burial customs of both the Shang and the succeeding Zhou periods and indicate the persistent presence of the dog in this context throughout the Chinese Bronze Age.

EASTERN SIBERIA

Beringia, the Pleistocene isthmus connecting Asia and North America, was apparently an open corridor during several time periods in which humans may have traversed the continental shelf exposed by the lowering of sea levels associated with the cooler episodes of the Ice Age (Colbert 1973b). Such meager evidence as is available suggests that humans entered North

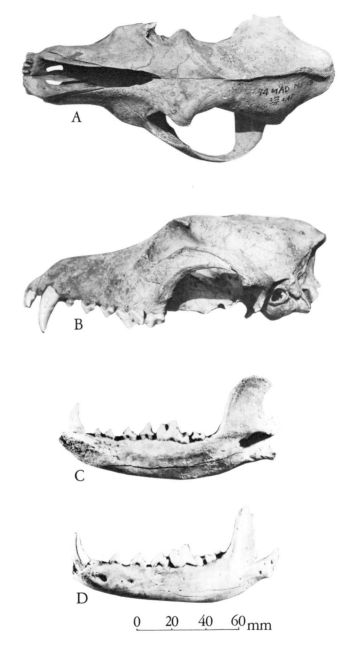

0 20 40 60 mm

Figure 5.8. Skull and mandibles of Neolithic Chinese dogs. A domestic dog, *Canis familiaris,* from Dadianzi, Inner Mongolia Autonomous Region, China: A) skull, dorsal aspect; B) skull, left lateral aspect. Domestic dogs, *Canis familiaris,* from Ashan, near Baotou, Inner Mongolia Autonomous Region, China: C) a right mandible, medial aspect; D) a left mandible, lateral aspect.

Figure 5.9. Skull and mandibles of domestic dogs, *Canis famil-iaris*, from the late Neolithic site of Qixiaying, Inner Mongolia Autonomous Region, China: A) a skull, dorsal aspect; B) same skull, palatal aspect; C) a left mandible, lateral aspect; D) a right mandible, medial aspect.

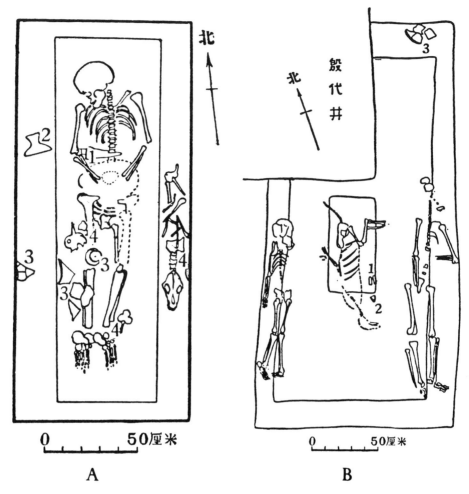

北

北 殷代井

0   50厘米

0   50厘米

A   B

Figure 5.10. Human (*Homo s. sapiens*) and domestic dog (*Canis familiaris*) burials from Bronze-Age Shang tombs in western section of Yin, Anyang, Henan (ca. 1384–1100 B.C.): A) human interment with accompanying dog placed on the *ercengtai* or "second-level platform," Tomb M656; B) human and canid sacri-

America from Eurasia near the end of the Würm (or Sartan, in Siberian terms) glacial period, although a disconcerting paucity of unequivocal evidence forces us to leave this chronological question unresolved.

The few early sites on the North American side of the Bering Strait with associated fossils that bear on the problem of the origins of dog domestication have been discussed in Chapters 2 and 3; but what of the early canids on the Siberian side of the Strait? Some of these finds should certainly help to bridge the gap in our knowledge linking the early Neolithic dogs of Asia with those in North America. As Powers (1973:14) pointed out, Pleistocene faunal

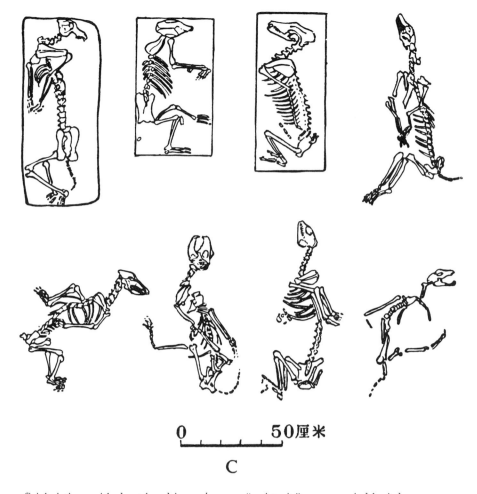

0   50厘米

C

ficial victims, with dog placed in *yaokeng,* or "waist pit," accompanied by jade plectra (1) and a bronze bell (2), Tomb M785; C) sacrificial dogs recovered from tomb *yaokeng* (boxed figures) and general pit fill. Note some individuals have had forelimbs bound behind back. From *Kaogu Xuebao* 1979(1):47–51.

assemblages are poorly known in the eastern U.S.S.R., outside of southern Siberia and Transbaikalia. Consequently, data presented here concerning the presence of the domestic dog in Siberian archaeological sites should not be taken as definitive evidence; rather, it should be borne in mind that while enormous strides have been taken in Siberian archaeology since the late 1950s, vast tracts of territory in the eastern U.S.S.R. were still virtually unknown archaeologically in the mid-1980s.

Siberia has produced enigmatic remains of large canids often referred to as dogs or domesticated wolves. The large Afontova Gora II site, near

Krasnoyarsk in the Yenisei river valley, was dated—on the basis of a single radiocarbon reading—at 20,900 ± 300 BP (GIN-117). Among the faunal remains collected at this site from 1923 to 1925 are bones "identified as dog or perhaps domestic wolf" (Chard 1974:30). Attempts to locate the original specimens upon which these determinations were made have proven fruitless; hence, the Afontova Gora II canid remains are mentioned here solely on the basis of their appearance in previous literature. As unsubstantiated finds, these canids must, for the time being, be considered anomalous.

Only when one approaches the Pleistocene-Holocene boundary does the Siberian archaeological record begin to yield substantial remains of domestic dogs. Dikov (1977), for example, reported on the discovery of apparently domestic dog remains at the stratified site of Ushki I in Kamchatka. Level 6 of this site (the middle of three stratigraphic horizons assigned to the Upper Paleolithic) has been radiocarbon dated at 10,360 BP to 10,760 BP (West 1980). However, although Dikov (1977:233) dismisses the possibility that ^{14}C dating errors could occur as a result of local volcanic activity, the true antiquity of Level 6, as well as of the other Paleolithic strata at Ushki Lake, remains unsatisfactorily dated. Of interest to our discussion here is that Level 6 has produced what the investigators believe is an intentional burial of a domestic dog, although the report on faunal remains contained in Dikov's (1977) report does not elucidate the criteria upon which this determination was made.

At Ust'-Belaia, in the Anadyr basin in the Siberian northeast, a prepared blade-core industry is found in association with such Neolithic components as ground stone tools, projectile points of several types, simple pottery, and occasional occurrences of bronze. Chard (1974:58) reports that a ritual dog burial, thought to be earlier than 9000 years old, is present in the site, and Dikov (1977) corroborates this find, although in 1984 the Neolithic occupation of Ust'-Belaia was believed to extend only to circa 3000 BP or perhaps a bit earlier. In his discussion of the preceramic "Mesolithic" occupation of Ust'-Belaia, Medvedev (1969:63) included the domestic dog in his list of "culinary debris" recovered in terminal Pleistocene–earliest Holocene horizons III and IV, which have been radiocarbon dated to 8960 ± 60 BP (GIN-96). Unfortunately, the absolute chronology of the Ust'-Belaia site was problematical (Powers 1973; Chard 1974:58), and the criteria upon which the domestic state of these canids was determined are unclear.

The site of Bel'kachi I is located on the middle course of the Aldan River, on the left bank near the mouth of the Ulakhan-El'ge River in the village of Bel'kachi, Aldan District, Yakut A.S.S.R. (Powers 1973:59). Mochanov reported that preceramic level VIII at Bel'kachi I, tentatively radiocarbon dated to 5900 ± 70 BP (LE-678) produced remains of dog or

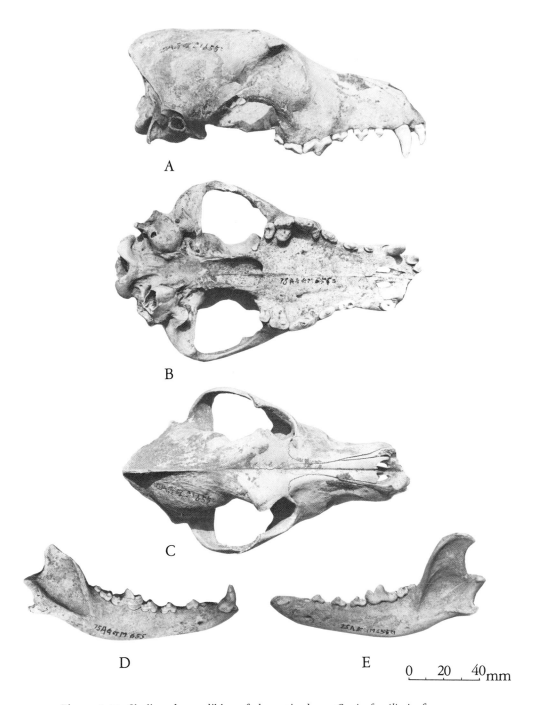

A

B

C

D   E   0   20   40 mm

Figure 5.11. Skull and mandibles of domestic dogs, *Canis familiaris,* from Bronze-Age Shang Tombs, western section of Yin, Anyang, Henan (ca. 1384–1100 B.C.): A) right lateral aspect of a skull; B) palatal aspect of same skull; C) dorsal aspect of same skull; D) a right mandible, lateral aspect; E) a left mandible, lateral aspect.

wolf which he classified only as *Canis* sp. (Mochanov cited in Powers 1973:61). Unfortunately, no detailed morphometric comparisons of these canid remains seem to have been conducted; hence, their specific taxonomic status remains unclear.

By the time developed Neolithic cultures appear in eastern Siberia, dogs are present in nearly all faunal assemblages reported. In the Lena river basin, the Kitoi and Glazkovo phases of the local Neolithic sequence are both characterized by separate dog burials (Okladnikov 1955:302) beginning about 2000 B.C. In the Vladivostok region of the Soviet Maritime Province, shell middens of the iron-using Sidemi culture, tentatively dated to the eleventh and twelfth centuries B.C., have yielded remains identified as those of domestic dog and pig (Chard 1974:94). These remains may indicate the beginnings of Chinese contact, through Manchuria, which is evident later in the prehistoric sequence.

In the 1980s the archaeology of eastern Siberia was only beginning to be understood in any detail at all. The critical late Pleistocene transition from gathering/hunting/fishing cultures to those of settled agriculturalists and pastoral nomads is one of the most poorly defined aspects of Siberian prehistory—one which is, unfortunately, essential to our understanding of the origins of canid domestication in northeast Asia. Because the pace of archaeological investigations being conducted in this region of the U.S.S.R. began to quicken in the 1970s and 1980s we can expect important new data relevant to this and other questions to emerge in due course.

PENINSULAR SOUTHEAST ASIA

As Higham, Kijingam, and Manly (1980:149) pointed out, "The literature on prehistoric dogs from mainland southeast Asia is almost silent...." In spite of the scarcity of domestic canid remains discovered in archaeological contexts on the southeast Asian peninsula, those which have been recovered suggest that this region contains data of great importance in the study of the domestic dog in the eastern Old World. Of particular importance is the fact that the indigenous canids of southeast Asia, the cuon (*Cuon alpinus*) and the golden jackal (*Canis aureus*), are apparently not the wild progenitors of early domestic dogs found in the region. In their analysis of domestic dog remains from a series of sites in northeast Thailand, Higham, Kijingam, and Manly (1980) demonstrated, on the basis of multivariate comparison, that these dogs have their closest morphological affinities with the wolf (*Canis lupus*), which is found no closer to the southeast Asian peninsula than southwestern China (*C. lupus chanco*) and the Indian subcontinent (*C. lupus pallipes*).

The best-documented of these Thailand settlements, Ban Chiang in Udon Thani Province, has produced occupation and burial layers ascribed to six prehistoric phases extending back to circa 3500 B.C. (Gorman and Charoenwongsa 1976). Even the earliest dog remains from Ban Chiang "...correspond with those from modern animals" (Higham, Kijingam, and Manly 1980:159). This fact suggests either that domestic dogs represent an intrusion into the southeast Asian archaeological sequence in a morphologically domesticated state or that still earlier evidence of dog husbandry remains to be found in this area. Arguing against the latter position is the fact that no canid bones have been found in the cultural debris of Hoabinhian archaeological sites, a culture presumed to have principally engaged in a broad-spectrum foraging economy beginning in the late Pleistocene. The canid remains from Ban Chiang bear evidence of cutting, breakage, and charring in prehistory, suggesting that dogs may have been commonly raised as items of consumption among Thailand's early rice agriculturalists.

The recognition of the importance of the southeast Asian tropical zone in the origins of horticultural and rice agricultural subsistence strategies in the Old World bears important implications on the question of early domestic canids in East Asia south of the Chinese-Vietnamese frontier. By 6000 to 5000 B.C., sites in the area of Hangzhou Bay, Zhejiang (i.e., Hemudu), and along the southeastern coast of China (i.e., Qingliangang) indicate that rice agriculture and the domestication of animals, including the dog, were firmly established.

At Non Nok Tha, also in northern Thailand, the excavation of a series of human interments extending back as far as 4000 B.C. revealed remains of several presumably domesticated animals, including the dog (Bayard 1972:15; Bellwood 1978:162).

In the Socialist Republic of Vietnam, early Neolithic (Bacsonian) contexts have yielded dog remains dating possibly as early as 5000 B.C. For example, at Da-But, 30 km inland from the sea near Thanh-Hoa, dog bones were reported from a marine shell midden in association with Bacsonian edge-ground tools (Patte 1932; Boriskovsky 1968–1971, Part VI).

At Phung-Nguyen in the Red River delta north of Hanoi, over 3800 square meters of a low mound have been excavated by Vietnamese archaeologists. The excavation has revealed a Neolithic assemblage which may date to as early as 3000 B.C. (Boriskovsky 1968–1971, Part VI; Nguyen 1975). In addition to rice grains and bones of pigs, cattle, and chickens, baked clay figurines of dogs have been recovered from Phung-Nguyen and other contemporary sites. Such items suggest a link between the appearance of rice agriculture and animal domestication, including *Canis,* on the southeast Asian peninsula.

69

Outside of Thailand and Vietnam, mainland southeast Asia has produced few domestic canid remains associated with archaeological complexes that clearly antedate the florescence of bronze metallurgy after circa 1500 B.C. In insular southeast Asia, there is little evidence to suggest the presence of the domestic dog prior to 2000 B.C. (Bellwood 1978:149), although the dingo, certainly a product of intentional human introduction, is present in Australia at possibly 4000 B.C. or earlier (Bellwood 1978:78; Barker and Macintosh 1979).

Given the fluctuations in mean sea level that characterized the Pleistocene Epoch, it is not surprising that a large sample of potentially crucial early Holocene archaeological assemblages have been, for all practical purposes, lost to science.

One critical subject for future research which bears obvious implications for the study of canid domestication in southeast Asia is the relatively early presence of the dingo in Australia. Perhaps derived originally from India, the dingo must have entered Australia through Indonesia, and yet the archaeological record remains silent on this point. This example is but one of many which illustrate the fact that, although the pace of archaeological activity in Southeast Asia has increased dramatically since the 1950s, by the mid-1980s major areas remained virtual blanks as far as our knowledge of early food-producing cultures was concerned (i.e., Kampuchea, Laos, Burma, parts of Indonesia and Malaysia).

Clearly, it would be premature to speculate on a topic as specific as the origins of the domestic dog in southeast Asia, given the obvious lacunae in our data base. All that can be stated at this point is that the known archaeological assemblages dating to the early Holocene in southeast Asia have not yet produced unequivocal evidence of the domestic dog, whose appearance in the region seems related to the influx of rice agriculturalists, perhaps prior to 5000 B.C.

6. Prehistoric Dogs in Europe and the Near East

Among the earliest domestic dogs in Europe are those from Senckenberg, Germany (Mertens 1936) and Star Carr in Yorkshire, England (Degerbol 1961). The dates of the two finds are compatible. The Star Carr dog has a radiocarbon date of 7538 ± 350 B.C. The Senckenberg dog seemed to be closer to a wolf, whereas the Star Carr dog appeared to be quite advanced according to Bökönyi (1974) and was not a recent offshoot of the wolves. This reasoning would suggest that domestication of the Star Carr dog would have, therefore, begun earlier. It was first thought that the skull fragments from Star Carr represented an immature wolf, *Canis lupus*. However, Degerbol (1961) used the proportions of a very short jaw and large teeth and overlapping premolars to determine to his satisfaction that this young individual was a true dog and not a wolf or a dog-wolf cross. J. G. D. Clark (1972) stressed that Degerbol's change of classification from the first questionable status of the canid from Star Carr to the positive determination of *Canis familiaris* was proven.

A cautionary note should be added in regard to assigning the Star Carr canid to a positive taxonomic category. The skull from Star Carr is quite fragmentary (Fig. 6.1). It is also of an immature individual and, therefore, due to the undiagnostic nature of immature canid bones, it is not a reliable indicator of a specific taxonomic category. The most positive statement that should be made in relation to this Star Carr canid is that it represents either *Canis* sp. (an undetermined species of *Canis*) or cf. *Canis familiaris* (perhaps a domestic dog). In fact, it may be more properly assigned to *Canis lupus familiaris*, a tamed wolf pup. It is not unreasonable to assume that if humans killed a mature mother wolf with pups some or one of the pups might be brought back to camp and allowed to mature.

Clark (1972) made a questionable assumption in regard to Pido-plichko's (1969) reference to the Mezin wolves. He stated that Pidoplichko reported some confirmation of the occurrence of dogs at the mammoth-hunter kill station at Mezin in the Ukraine. Actually, Pidoplichko merely noted that some of these wolves had shortened faces and perhaps represented the initial stages of taming the wild animals. By placing them in the taxonomic category of *Canis lupus domesticus*, he merely likened them to tamed wolves, not domestic dogs.

Crania of dogs from the Mesolithic of Sweden and Denmark have been dated from 6800 to 5000 B.C. Some of these may have constituted food items if we are to judge by the broken neurocrania (the bones may have been broken to extract the brain). Some knife-cut marks are visible on the surface of the bones, which may be interpreted as evidence of butchering. The dogs of Denmark appeared to be shorter-limbed but otherwise quite similar to the Star Carr canid.

Rather extensive reports of later dog finds in Europe have been given by Epstein (1971) and Bökönyi (1974), but these need not be repeated here, since they deal with already well-advanced domestic dogs and have little or no bearing on the early development of the domestic dog in its earliest stages.

It is interesting to note that several European specialists on the origins of the domestic dog are hesitant to accept evidence of early dogs in North America. Epstein (1971), referring to the Jaguar Cave dogs and the Pleistocene canid from Illinois, stated that "there cannot therefore be any doubt about the domesticated state of these dogs, although the dating of their remains may seem exceedingly early for American dogs." Bökönyi (1974) also stated: "According to our present evidence the earliest remains of domestic dogs were found—against all expectations—in North America." Why such statements were made, considering the logical possibility of such finds being made because of the ancient direct passage that was available via the Bering Strait from Asia to North America, is somewhat surprising.

As of 1984, the oldest reported domestic dog in the world was from the archaeological site of Palegawra Cave, located in northeastern Iraq. An estimated age of 12,000 B.P. has been assigned to this dog (Turnbull and Reed 1974). The record of this earliest find of *Canis familiaris* is unfortunately based on a single, small, mandibular fragment consisting of the main body of a left horizontal ramus that lacks the symphysis angle, and ascending ramus. The partial dentition consists only of $P_{\overline{3-4}}$ and $M_{\overline{1-2}}$ and alveoli for C and $P_{\overline{2}}$ (Fig. 6.2). There seems to be little doubt among the osteologists who examined this jaw fragment that its morphological shape and character is closest to the similar-sized Kurdish domestic dogs and the prehistoric dogs from Jarmo with which it was compared. However, it is certainly within the realm of

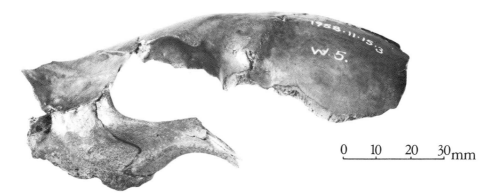

Figure 6.1. Fragmentary skull of immature canid from Star Carr, England, left lateral aspect.

possibility, and to be considered, that it may represent an aberrant wild canid, since it exhibits some wolf characteristics, as well as those determined to be typical for domestic dogs. If additional fragments of similar-sized and similar-structured individuals—even if no more complete than the type fossil (PM 11265)—were to be recovered from the Palegawra Cave site, it would help to dispel the doubts as to whether or not this specimen has been assigned to the correct taxonomic category and should indeed be listed as *Canis familiaris* or, if not, perhaps as *Canis lupus familiaris*.

There is a problem associated with fossil dogs of the Near East that is not unique to that area. It is simply that adequate associated post-cranial bones are rarely found or collected with the skulls or mandibles. In the past very little attention was paid to non-human animal bones, and some of this lack of post-cranial evidence in the case of dogs may possibly be due to selective collecting on the part of the excavators. It is true, of course, that the preservation of some skulls may be due to special burial practices that involved only these elements. However, a great number of Chinese finds have recovered entire skeletons from numerous dog burials in association with humans. One such occurrence, discussed in Chapter 5, is at the Bronze-Age, late Shang tombs in the western section of Yin Xü at Anyang, Henan (see Fig. 5.10), where hundreds of the tombs contained dog burials, many associated with people (Anyang Archaeological Team, 1979). Happily, during the 1970s and 1980s there was a noticeable increase by archaeologists to carefully collect and save most of the non-human, as well as human, bones that were encountered in excavations.

The Near East, like China, has a wolf among its fauna. Named *Canis lupus pallipes*, it is smaller than the majority of wolves found in other parts of the world, but comparable in size to the Chinese wolf, *Canis lupus chanco*. One difficulty of conducting adequate comparisons of excavated canids from Near Eastern sites is the lack of large comparative collections that contain adequate skeletons of Near Eastern wolves. I have found that no single museum has adequate collections of skulls or mandibles with which to undertake an acceptable multivariate analysis. Due to the decreasing numbers of truly wild canids the world over, I see no solution to this problem in the future. It is no longer possible to amass large collections of skeletons of the rarer living animals, and this is as it should be. However, this lack of both archaeological canid material and recent osteological material at times results in several changes of opinion concerning the same site and, at times, the same specimen, depending on who is doing the analysis. This situation is fully understandable to anyone who has had to face this problem and isn't satisfied with the taxonomic listing of *Canis* sp. for a bone or bones that could be either a domestic dog or a local wolf.

Some (perhaps the majority of) classifications of early *Canis familiaris* that were based on very fragmentary material were really just clutching at straws in an attempt to establish the presence of the earliest domestic dog. There seems to be competition among faunal workers for the "first" domestic canid from sites the world over. This apparent haste to record a new, and older, domestic dog has sometimes led to uncertainty regarding the taxonomic assignment of dog remains and to the use of zoomorphic representations, interpreted as "dogs," to bolster the scanty skeletal evidence that was being used. This problem is quite apparent in a number of published citations of early dogs that appear to have been based on less than adequate skeletal evidence. For example, in a report that relates to the beginning of animal and plant domestication in southwestern Iran (8000 to 9000 B.C.), Hole, Flannery, and Neely (1969) discussed the recovery of canid remains. They stated that osteological evidence for the presence of the domestic dog at Khuzistan, Iran, was based mainly on the increase of *Canis* remains in the Sabz Phase (6000–5500 B.C.) of the excavations. They noted that none of the canid remains showed evidence of having been used as food; by contrast there was evidence of such use during the Mohammad Jaffar Phase (6000–5600 B.C.). Some tooth crowding, like that which is normal in similar-sized domestic dogs, was noted in the Khuzistan canids. These animals were about the size of Kurdish guard dogs and similar in size to the more familiar Irish setter (no taxonomic association with the latter is implied). The authors concluded that this evidence probably suggested that the domestic dog was present in

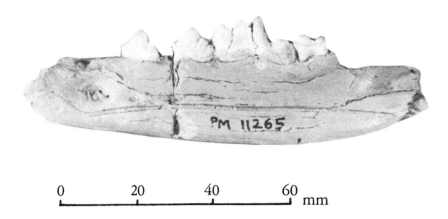

Figure 6.2. Fragmentary canid mandible from Palegawra, Iraq, left lateral aspect.

Khuzistan by 5500 B.C. and that these dogs were probably descendents of the local wild wolf, *Canis lupus pallipes*. The writers mentioned the difficulty of separating the bones of this wolf from those of the domestic dog, due to the small size of the wolf and the wide range of its osteological variation. Hole and his colleagues (1969:314) stated correctly that "representations in prehistoric art are always dangerous evidence on which to base zoological conclusions." However, they then depicted sherds decorated with zoomorphic representations and referred to them as evidence for the appearance of the early dogs of Khuzistan.

Charles Reed is certainly one of the most respected faunal analysts who has worked on the topic of early animal domestication in the Near East, but even a person with his background has, at times, used questionable evidence to establish early domestication dates. For example, he used ceramic figurines at Jarmo (6700 B.C.) to aid in establishing the presence of domestic dogs there: he first regarded the canid remains at Jarmo as representing wolves and reported that dogs were not present in the midden (Reed 1959); later, he stated that the best evidence for the domestic dog at Jarmo was not biological, but cultural, and was based on a figurine that he interpreted as a dog-like animal (Reed 1960). However, he was still puzzled, and rightly so, by the absence of any skeletal remains of domestic dogs. A year later he shelved the evidence based on clay figurines, and a skull, which he had tentatively identified as a dog, was reinterpreted as belonging to the wolf of southwestern Asia, *Canis lupus pallipes* (Reed 1961). After a thorough reevaluation of the faunal remains several years later, the large canids from Jarmo were removed from the category of wolves, as "they undoubtedly represent dogs not wolves" (Reed 1969:370). Then, after an intervening decade, all of the recovered canids from Jarmo were subjected to a special study (Lawrence and Reed, in press) in order to determine the status of the domestic dog in the Near East during the period 9000 to 8500 B.P. Of the fifty-three cranial and mandibular fragments of *Canis* that were examined, eighteen were positively identified as *Canis familiaris*, and none was clearly determined to represent *Canis lupus pallipes*.

Comparisons of identified dog crania in museum collections with those of Near Eastern wolves have borne out the difficulty of separating either species, dog or wolf, on fragmentary material. The results of a special study by Lawrence and Reed (in press) strongly suggest that the ancestral form of the Jarmo domestic dogs was a local race of the Near Eastern wolf. However, whether this was *Canis lupus pallipes* or *Canis lupus arabs* or another as yet unestablished race is yet to be determined. It is important to point out that the foregoing changes of opinion regarding the specific assignment of the Jarmo canids are fully understandable to anyone who has worked with the archaeological material from this family of carnivores for any length of time. However, I believe that zoomorphic representations, which can be interpreted in many ways by many people, should not be used as supporting evidence of early dog domestication when the skeletal material is inadequate for that purpose.

The fauna from the pottery and pre-pottery Neolithic levels of the tell of Jericho included some interesting canid remains. Zeuner (1958) examined the canid teeth from these levels and stated (p. 53) that "the shape of these teeth leaves no doubt that they belong to *Canis*. They might come from a small wolf, a jackal, or a dog." He opted for the dog, based on what he termed "slender evidence only, namely on size." He ended his discussion by claiming that the great variability in the size of these teeth suggested conditions of domestication and the presence of more than one breed. Clutton-Brock (1969), who also assessed the composition of the faunal assemblage from Jericho, arrived at an entirely different conclusion regarding the canids that were encountered and recovered. She stated (p. 338) that "contrary to Zeuner's statement, careful measurement of all the Jericho specimens does not show any great diversity in size of bones and teeth." She goes on to say that measurements of the bones of *Canis* spp. from Jericho give

> ....no indication of which species are present nor of...evidence for domestication....Osteological characters cannot provide evidence for distinguishing tamed wolf from early

domestic dog, and it is in fact still doubtful how clear are the distinctions between wild and tame wolves. But it is of course possible to distinguish wild wolf from fully domesticated dog. There is no evidence for such a dog from any of the Jericho levels.

Clutton-Brock is recognized as a most thorough and careful researcher in the field of faunal determinations. I agree with her conclusions, and I hope that other workers will follow her lead and exercise a bit more caution in assigning the name *Canis familiaris* to every scrap of wild canid that exhibits any individual variation from what they consider the norm for this species.

Lawrence (1967) briefly reported on a domestic dog from south-central Turkey having ^{14}C dates that cluster around 7000 B.C. The site is named Cayönü and is near Ergani in the Province of Diyarbakir. The site was excavated in 1964 by a joint archaeological team of the University of Istanbul and the University of Chicago Prehistoric Project. The dog remains that were found consist of a pair of rami of a rather small individual about the size of a large Pueblo Indian Dog (see Fig. 3.4), but more heavily structured. The jaws are thickened latero-medially, and the teeth are comparatively large and crowded in the tooth row. Indications, projected from these incomplete elements, suggest a short-faced dog. All characteristics indicate that it is, indeed, *Canis familiaris* and not *Canis lupus*. As with other comparable, incomplete finds, I hope that additional material will be collected from Cayönü in the future to support these first conclusions. Lawrence (1956) also examined canid bones from a sink-hole south of Haditha on the Euphrates, collected by a Peabody Museum of Archaeology Expedition (Harvard University) in 1950. These bones appeared to be similar in comparison to a moderately large domestic dog. Lawrence also compared them with a series of Mesopotamian wolves, *Canis lupus pallipes*, and found that the wolf skulls showed some overlap in morphological characteristics with native domestic dogs in those features customarily found to be reliable in separating the northern races of *Canis lupus* from those of *Canis familiaris*. She concluded that the two canid skulls from the excavations, if taken by themselves, seemed to be more like dogs than wolves. Again, this would have to be considered a conclusion based on a feeling rather than a fact.

Meadow (1983) reported the occurrence of an early domestic dog from an archaeological site at Hajji Firuz in northwestern Iran. Radiocarbon age determinations range between 5500 and 5100 B.C. This canid was determined from four cranial fragments and one vertebra, but Meadow stated (p. 6): "The cranial fragments and teeth surely come from dog and not wolf (*Canis lupus*) nor jackal (*Canis aureus*)." As with some other identifications of near Eastern domestic dogs, perhaps this one should await additional supporting material before it is unconditionally accepted. Given the small number and nature of the fragments, it is still within the realm of possibility that an aberrant individual of a wild species is represented.

An early drawing or outline of an animal represented in a hunting scene (on a ceramic vessel) has been interpreted as being representative of a domestic dog. This zoomorphic representation was from level III (c. 7500 B.P.) at Catal Hüyük in south central Anatolia. Reed (1969:371) stated that the "dog" would probably not have been recognized as such if it were not for the context of the scene of a man hunting deer, because the actual figure of the small carnivore could easily be that of a mustelid or viverrid. This pictorial representation, which may represent any one of a number of non-canids, cannot be taken as evidence of dog domestication at Catal Hüyük. In a brief report on this site, Mellaart (1964:5) stated that "dogs also appear to have been domesticated," but he gives no supporting evidence, skeletal or otherwise. Presumably, he was using the questionable drawing referred to by Reed. For this, and for most of the Near East canid finds assigned without question to *Canis familiaris*, the supporting data are too fragmentary and inadequate to form a basis on which to support such a taxonomic listing with any degree of certainty.

7. Canids of Uncertain Taxonomic Status

Some taxonomic listings for dogs, discussed in earlier chapters of this book, have generally been accepted even though they were based on fragmentary evidence. There are other canids whose taxonomic status is based on evidence that is too shaky to allow to go unchallenged. This second group differs from the first in that the stratigraphic occurrences of the specimens are highly suspect. In all instances the levels from which the bones were reported may have suggested an older age (in the literature) than they are in reality. Most are quite possibly intrusive forms from a higher and more recent level to the lower and older level in which they were found. In some instances animals not fully mature were used to establish the presence of domestic dogs, but such species identification is generally not possible from fragmentary, immature bones and teeth.

There have been a number of published references to the occurrence of small domestic dogs from the Pleistocene beds of North America. Some of these determinations were based on single elements and all were from strata that could easily represent a re-deposition or intrusion. I feel that they should be listed as problematical, and not as unequivocal, finds from known stratigraphic contexts. Aside from the questionable stratigraphic occurrences of these reported dogs is the fact that, since they are each representative of a single individual, one must take into account the possibility that they might represent aberrant variations, both in size and form, of a wild species of *Canis*. They are discussed in this book for possible reevaluation if additional specimens are encountered in the future.

Galbreath (1938) reported a left mandible of a domestic dog from a gravel pit in east-central Illinois which dates to the Wisconsin age of the late Pleistocene (Fig. 7.1). Due to the fact that the gravels were worked with a power

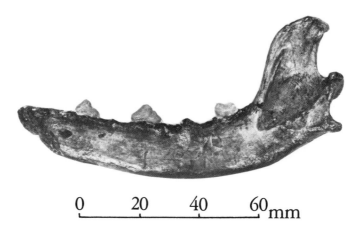

Figure 7.1. Canid mandible recovered from Illinois Pleistocene gravel bed, left lateral aspect.

shovel, the age of a number of the vertebrates had to be designated as "uncertain." Galbreath stated that some of these may have been mixed in the strata by the excavating equipment. Galbreath (1947) published a short addition to his original paper, but no vertebrates were included. In the mid-1980s, the jaw of this small canid remained the single, isolated find recorded from the Pleistocene deposits of Illinois. I believe that it should be given questionable status, based on the circumstances surrounding its recovery, because it is from a worked gravel bed and also because of its small size and advanced characteristics from a stratum of this early age.

Stock (1938) reported on a coyote-like wolf jaw from the Pleistocene tar pits of Rancho La Brea in southern California (Fig. 7.2). The comparisons that Stock made convinced him that the animal was too small to represent a wolf and was closer to a coyote. Due to the differences observed, including a heavier and shorter mandible than that found in recent coyotes, he coined a new specific name, *petrolei*. Periodically, this specimen is considered as a candidate for establishing a new record of a Pleistocene dog from the Rancho La Brea tar pits. As of early 1985, none of these descriptions has gone beyond the manuscript stage. When this jaw is compared to the coyote, there is still a question as to the possibility of its being an aberrant, wild canid. A still greater question arises when one considers the fluid consistency of the tar pits. Is this canid jaw really contemporary with the known extinct forms from La Brea, such as the mastodon, mammoth, and saber-tooth cats, or is it an

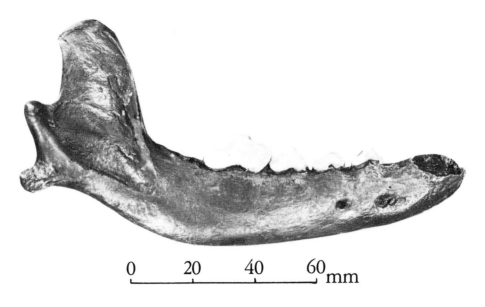

Figure 7.2. Mandible of canid from Rancho La Brea, California, right lateral aspect.

intrusive bone that has found its way down from a later period of time to finally rest with the older Pleistocene bones? Without additional material or evidence, it must either be considered as coyote, as described by Stock (1938), or be tabulated simply as *Canis* sp.

In 1942 the mummified remains of a small, immature dog, along with some isolated bones of other individuals, were recovered from Ventana Cave on the Papago Indian Reservation in southern Arizona (Figs. 7.3 and 7.4). This rock-shelter was occupied on numerous occasions for varying lengths of time from the late Pleistocene up until comparatively recent times. The site report (Haury 1950) listed a mixed fauna of extinct, as well as living, animals that had accumulated in the floor fill of the cavern. More recently, Colbert (1973a) described an extinct horse which had an associated ^{14}C date of 11,300 to 12,000 B.P. Colbert believed that this horse was contemporaneous with the human occupants of the cave at that time. However, the dating of the dog remains is still quite controversial. Haury (1950:158) stated: "The cultural association of these dog remains was Chiricahua-Amargosa II complex, and in terms of the Christian calendar, the age may be placed some millennia before the time of Christ. If we follow Antev's dating of the Cochise culture, this would have been between 800 and 300 B.C. or after 2500 B.C. if we follow Bryan." He goes on to state (p. 159), in reference to the dog mummy in particular: "The mummified dog mentioned above probably lived less than a thousand years ago and was doubtless kept by members of the Hohokam

81

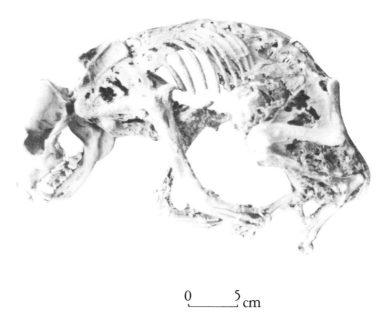

0   5 cm

Figure 7.3. Mummified remains of a small, immature domestic dog, *Canis familiaris*, from Ventana Cave, Arizona.

culture." In view of the broad range of possible dates for the Ventana Cave dogs, they are included here with the other questionable canids, even though Ventana Cave itself is one of the oldest sites in the Western Hemisphere that was occupied by humans.

Underwater deposits in the springs, sinkholes, and rivers of Florida are most unreliable as far as interpreting any stratigraphic association of bones which occur and are recovered in these areas. It can be most misleading to state that a bone has "the same preservation as most other material" (Webb 1974:127) in relation to distinguishing a late Pleistocene bone from one that is of more recent origin. Such a statement is subject to qualitative individual observation if it is not substantiated by a defined analysis, test, or some other accepted standard of determination. I have spent many years collecting Pleistocene, as well as recent, animal bones from the inland waters of Florida and have found that the weight, color, or even texture of recent bones can be altered considerably and rapidly by the amount of minerals in the water, which varies from one stream to another. A heavy concentration of sulphur, such as in the Itchetucknee River, will color the bone of a modern pig (*Sus scrofa*) so that it has the same appearance as that of a Pleistocene tapir (*Tapirus veroensis*). The clay beds in this particular stream contain both

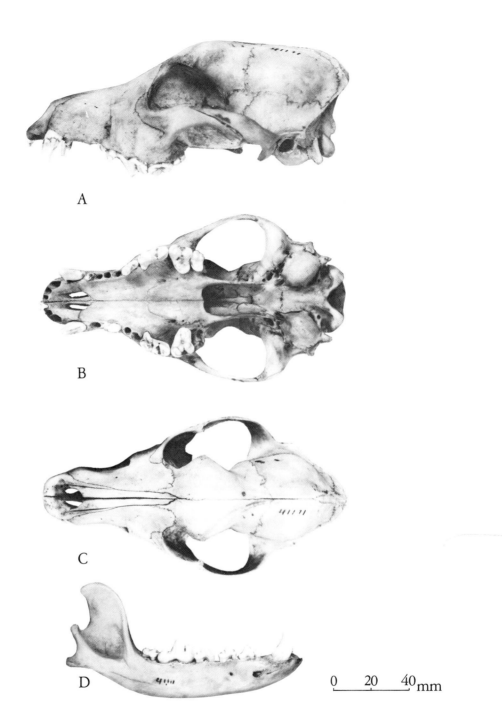

A

B

C

D

0   20   40 mm

Figure 7.4. Skull and mandible of a domestic dog, *Canis familiaris,* from Ventana Cave, Arizona. A) left lateral aspect of skull; B) palatal aspect of skull; C) dorsal aspect of skull; D) lateral aspect of right mandible.

Pleistocene and recent animal bones which protrude, to one degree or another, into the swift, running waters. This mixing of bones is due to redeposition, which goes on continually. The Pleistocene bones are all black or dark brown in color when removed from the water. The recent bones remain almost white or buff colored on the portions that are completely embedded in the clay, while those areas that are exposed to the percolating, mineral-laden water become almost as dark as the Pleistocene fossils.

There is a well-known fossil locality in central Florida, a sinkhole known as Devil's Den, which was a natural trap for animals that fell into the hole and drowned in the water which filled the lower portion. The trapped animals that have been identified range in size from shrews (*Blarina*) to the largest of Pleistocene ground sloths (*Megalonyx*); there are several humans, as well. The age of this fauna encompasses the very late Pleistocene and Holocene and includes animals that were victims as recently as the present. Webb (1974) listed a find of *Canis familiaris* from this locality consisting of five mandibular rami, a right maxilla, a cranium, and some other skull and limb fragments. He identified these canid bones as being the size of a present-day coyote, and he believes this find represents a domestic dog from the late Wisconsin period of the Pleistocene. He stated (p. 128): "The association of these dogs with man can almost be taken for granted." However, this statement is not supported by convincing evidence. What is disturbing about the vertebrates at this site is the occurrence of bones of domestic pigs, *Sus scrofa*, domestic cattle, *Bos taurus*, and horses, *Equus caballus*. The genus *Equus* also occurred in Florida up to the close of the Pleistocene, and species of this age cannot be readily separated from the modern horse on fragmentary evidence. The other two animals, pigs and cattle, were not known in the Western Hemisphere before their introduction by the Spanish in the early 16th century. Webb (1974:138) apparently had some doubts about his positive identification of the domestic dog, which was of small size, to the late Pleistocene of Florida, because he stated: "Although they [the canid bones] have been recovered only from the surface layer of the deposition, extinct animals also occur in that layer. This suggests to me, but does not prove, that the *Canis familiaris* material accumulated with the rest of the mammalian fauna at Devil's Den during a late Pleistocene interval." The location of the bones does seem to exclude the possibility that the dog is no older than the remains of the domestic pigs and cattle. On the other hand, it is a certainty that bones and other heavy intrusive objects can be subject to a certain amount of mixing when deposited in water-covered sediments. It is, therefore, equally certain that the exact stratigraphic position of such objects in relation to one another can never be accurately ascertained.

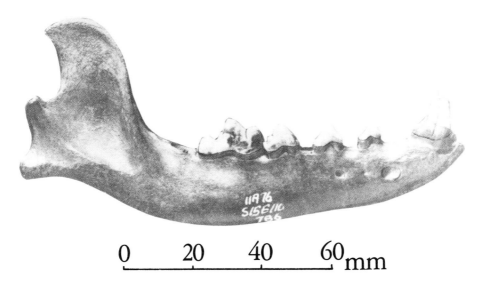

0   20   40   60mm

Figure 7.5. Mandible of canid from Old Crow, Yukon Territory, Canada, right lateral aspect.

Beebe (1978, 1980) reported a new find of a small domestic dog mandible from Pleistocene beds of the pre–late Wisconsin period in the Old Crow Basin, Yukon Territory, Canada (Fig. 7.5). She stressed that the domestic status of this canid and its stratigraphic position are "probable," not certain, and that the highly evolved morphology of this small animal suggests that it represents a rather late stage in the process of domestication. Canby (1979:348) quotes Beebe, in regard to this find and some additional fragments collected a bit later, as referring to "...the jaws of several domesticated dogs, some of which appear to be at least 30,000 years old. This is almost 20,000 years older than any other known domesticated animal anywhere in the world." Canby goes on to voice his own opinion, which is held by many others, including the present writer, that "the problem of accurate dating casts a shadow across the dramatic discoveries at Old Crow." None of the canid finds has been recovered in unquestionable association with datable human traces, such as hearths or charcoal.

Redeposition of bones and artifacts in arctic soils, due to solifluction, is a recognizable problem. Rainey (1939:391) stated:

The "flow" of soil from upper slopes down into broad glacial
stream valleys, brought about by alternating freezing and

thawing (a condition which can be observed at the present time in central Alaska), may prove to be an explanation; if so, it is probable that some vegetable as well as faunal remains in the muck have been redeposited, or moved down the slopes during many different advances.

If the Old Crow canids represent examples of animals that were tamed from local wolves, and if the dates given for their occurrence are accurate, then they should be considerably larger. On the other hand, if they are the progeny of an older, Asian, ancestral pool, then the size is about what should be expected. Until such time as more corroborating evidence is produced that links the Old Crow canids firmly to an early date, it is best to place them in the same category with the reported dogs from the Pleistocene of Illinois and Florida.

Areas in other parts of the world have also yielded problematic canids. Davis and Valla (1978) reported on a domestic dog burial from the Natufian of Israel with estimated dates of 11,310 ± 570 B.P. and 11,740 ± 740 B.P. They stated that the fragmented skeleton of a puppy was either a dog or a wolf. The crowding of the tooth row was given as the primary evidence used to support their case for its being a domestic dog. However, in a five-month-old dog or wolf of the same size, based on one fragmentary specimen, this characteristic cannot be accepted as conclusive evidence to support the presence of an early domestic dog. As with the Star Carr puppy from England (Clark 1954; see Fig. 6.1), the best that can be stated is that it may represent an early example of wolf taming; thus, it should be listed as *Canis lupus familiaris* or perhaps simply *Canis* sp., to indicate it is either a tamed wolf or possibly a domestic dog, but not conclusively either. Concerning this specimen, Davis and Valla (1978:609) stated: "The puppy, unique among Natufian burials, offers proof that an affectionate rather than gastronomic relationship existed between it and the buried person, an addition to our knowledge of the way of life of Natufian hunter-gatherers." Do we then postulate, using this same line of logic, that since some graves of Paleolithic hunters in Europe have been discovered with the scapulae of mammoths, *Mammuthus primigenius*, overlying the human burial that this physical proximity suggests an affectionate relationship between Pleistocene proboscideans and humans? The possibly erroneous assumption made by Davis and Valla was in stating that there were only two possible relationships between humans and canids— "affectionate or gastronomic." Why not religious, ceremonial, or some hunting fetish relating to the wolf and humans? In addition, no real consideration was given to the possibility that this case might be an occurrence of an aberrant wild individual, since only one fragmented individual was used as evidence. What do we know of the range of variation in five-week-old Natufian wolf skeletons or of the variation between young male or female animals?

The origin of the dingo is very controversial, and perhaps no canid has been given a more diversified taxonomic classification. It has been listed as *Canis dingo* by numerous taxonomists (Ewer 1973; Mech 1970; and others). More conservative workers have included the dingo under the same classification as the domestic dog, *Canis familiaris*, but with the subspecific taxonomic designation of *dingo* (Jones 1921). Van Gelder (1978) and Clutton-Brock et al. (1976) regard the dingo as a distinctive, feral, domestic dog. Clutton-Brock et al. (1976:145) stated that "it is not a biological species but a feral dog closely related to pariah dogs." They resolved the issue by stating (p. 141–142), in regard to the dingo and other domestic dogs, that "these two dogs have been treated as a separate species on an equal level with the wild species." They simply list both the dingo and other domestic dogs as "*Canis* (domestic)." Clutton-Brock et al. also stated that formal zoological nomenclature should be avoided in naming domestic animals. The present writer disagrees, since morphological differences between wild and domestic animals of the same genus are as great as that between different wild species of the same genus. If anything, eliminating domestic species would only further confuse an already complicated taxonomy of both wild and domestic animals.

The unsolved point of controversy concerning the dingo is just where this animal fits into the picture of canid domestication or whether it is even to be considered as a form of domestic canid. The ancestry and origins of the dingo have not been satisfactorily explained and the published evidence is not at all conclusive. Fox (1978:249) cited a date of 10,000 to 8000 B.C. for the first recovered dingo remains, but he did not elaborate upon this statement. Rine (1965) postulated that the Australian dingo came to that continent from China over a land bridge that connected Australia with Asia some thousands of years ago, but, again, no supporting data are volunteered to support this hypothesis. Jones (1921) postulated that the dingo was brought by sea by the first aboriginal colonists. A mummified dingo (Fig. 7.6), now in the Western Australian Museum, was collected by a survey group in a cave on the Nullarbor Plain. It was not associated with cultural remains and was dated on soft tissue to 2200 B.P. (Gollan 1977). A considerable number of dingo skeletons have been collected in Australia, yielding dates of about 3000 B.P.; older dates of 8500 B.P. have been attached to isolated teeth believed to be dingo, but this evidence is too meager to allow definitive interpretation.

There is considerable evidence pointing to failures by humans to successfully domesticate a non-hybridized dingo. Workers who have considerable experience with the training of police dogs have failed to produce anything resembling obedience in a dingo. Barker and Macintosh (1979) have compiled a review of the literature relating to the dingo and the problems that exist relating to its position as a domestic dog.

Aborigines in the heart of the Gibson Desert in western Australia have been observed in their relationship with the dingo. The animals were handled and touched by the aborigines, but rarely fed. They existed on what they could procure on their own. At the beginning of a hunt they were driven away from the hunting party, as they were considered to be more of a hindrance than a help in obtaining game. Why then would they be tolerated in an aboriginal camp? The answer to this seems to be their use as a means of warmth during near freezing nights in the desert: aborigines and dingos huddled together for mutual warmth (Fox 1975). The dingo remains an enigma whose true place in canid taxonomy will be clarified only by the recovery and analysis of additional specimens that are derived from archaeological contexts.

There has long been controversy over the size differences that may exist between the ancestral forms of Pleistocene *Canis* and the earliest animals that are considered to be *Canis familiaris*. Haag (1948) stated in his summary on an osteometric analysis of some aboriginal North American dogs that there is "a marked tendency of the size of the dogs found in archaeological horizons [to change] with the temporal position of that horizon, to wit, the older the horizon the smaller the dog." My observations, based on comparisons made over thirty-five years on a number of canid skulls and jaws, differ from this statement made by Haag. He did not have any of the Ukranian specimens (see Figure 2.1) available for study which suggest wolf taming in the late Paleolithic, although the Fairbanks, Alaska, wolf skulls (see Figures 2.2, 2.3) were available for comparison in the Frick Collections at the American Museum of Natural History in New York City when he examined the Eskimo dogs in the collections there.

Lawrence (1968) has made an interesting observation in relation to the size of prehistoric dogs in North America. She stated, in reference to the Jaguar Cave domestic dogs:

> The specimens originally reported (Lawrence 1967) are animals not very different in size from the small Basketmaker dogs of the Southwest. This had seemed to confirm the theory that there has been an increase in size of Indian dogs from small animals associated with the earliest cultures to larger, more recent forms (Haag 1948). Now, with the discovery of large dogs from the same levels in Jaguar Cave as the

Figure 7.6. Mummified remains of a dingo from Thylacine Hole, Nullarbor plain, western Australia (2200 ± 96 B.P.). Photo by Klim
Gollan.

small ones, we have evidence that smallness is not necessarily
related to the antiquity of the dog population. Actually, variation in size is one of the results of domestication of the dog,
and the pronounced difference between these specimens
indicates that this aspect of domestication is an ancient one.

It seems logical to assume that undiscovered forms of domestic dogs will
connect the known forms of ancestral *Canis lupus* with the rather adequate
representation of *Canis familiaris* from the early Neolithic and that these
forms will be of a size intermediate between the larger wolves and the
smaller dogs.

appeared in print and that were based on supposedly valid taxonomic characteristics. The Star Carr canid has been identified as a domestic dog only on the evidence of the crowding and displacement of the teeth, since this characteristic is known to be present in the domestic dog but is regarded as absent in wolves. However, it is not known whether or not, or how frequently, this condition may occur as an abnormality in wild European wolves.

In the George C. Page Museum at the Rancho La Brea Tar Pits in Los Angeles, California, there is a striking exhibit of 404 skulls of dire wolves, *Canis dirus*, all exhibited together on a translucent, backlit, wall panel. Here one can view and compare a considerable number of skulls of an extinct canid with little effort. These giant wolves were in no way influenced by human activities; there is no evidence of any association between these two groups of animals. However, it is interesting to note that about 25 of the specimens exhibit the characteristic "dished" configuration of the rostral area, anterior to the orbits, which is generally attributed to the domestic dog but not the wolf. This fact points up the folly of using any *single* osteological characteristic or of using fragmentary canid elements found to push back dates or to establish the presence of domestic dogs at early archaeological sites.

Responsibility for some of the error and confusion that exists in assigning positive, specific designations based on fragmentary bones or teeth must be borne by the vertebrate paleontologists. Until around the 1960s a common fault of paleontology was (and to a minor degree still was in the 1980s) to establish a new species taxon for mammals, based, at times, on a single tooth or on a fragmentary jaw, while more or less ignoring the possibility that these incomplete specimens might well be variants or aberrant individuals of already well-known and recorded species.

Anyone who has compared *Canis* skulls for any length of time has noticed obvious features that seem to separate wolves, coyotes, and domestic dogs. These differences are more or less of a gross nature, such as the swelling in the frontal area in dogs, comparatively smaller dentitions overall, and the development and degree of the overhang of the sagittal and occipital crests. Some of these changes from wolf to dog have been interpreted as being caused by the process of domestication. Radinsky (1973) discussed the cranial expansion, or swelling, so apparent in the domestic dog, as being, of course, a reflection of brain development in the frontal area. He mentioned that the prefrontal cortex of the brain—comprising the orbital and prorean gyri—functions in an inhibitory capacity with respect to behavioral patterns. Thus, its expansion might indicate more flexible or sophisticated behavioral patterns. There seems to be evidence that this relatively larger prorean gyri is related to all canids with a pack social structure. The evidence does not, however, explain why this area is noticeably larger in the domestic dog than in the wolf, both of which are pack animals, unless the additional change from wolf to domestic dog was an influencing factor in this development.

As with most archaeological material, specimens generally aren't complete enough to take all of the measurements required or desired. With complete skulls and mandibles it is possible to obtain 75 measurements that have been determined to be of possible morphological value. Of these, 13 essential measurements were selected for multivariate analysis (Fig. 8.1). All were within the realm of individual variation if used alone and were considered not to be reliable as single characteristics to separate the various species of *Canis*. I have chosen not to list the numerous pages of columns of measurements that would be required to present all of my data for determining the various species of *Canis* discussed here. Instead, all of this information, stored on both computer keypunch card decks and on computer discs, is available through the Arizona State Museum, University of Arizona, Tucson.

I have encouraged my colleagues to use the outline drawings of canid skulls and jaws illustrated by Von den Driesch (1976) when referring to comparative osteological measurements so that in the future some method of standardization will result in regard to the points from which measurements are taken. Over the years I have often checked measurements of certain areas of the skull or mandibles that have been published by various research workers. Using the same individual skulls as listed by Nowak (1979) and others, in order to check their measurements, I found that the published measurements often varied from those that I took. This variation depended on where the points of the measuring caliper were placed. On a wavy suture line this placement could result in a difference of several millimeters. The same was true for tooth measurements, depending on whether the caliper was placed above, below, or on the cingulum of the tooth being examined.

The photographs and accompanying millimeter scales used in this book were designed so that any interested scholar might take most comparative measurements directly from the published figures. The critical skull areas selected for measurement and analysis are as follows:

Skull length	Measurement 1
Cranial length	Measurement 2
Nasal / premax. length	Measurement 3
Premax / orbit length	Measurement 4
Tooth row length	Measurement 5
Molar length	Measurement 6

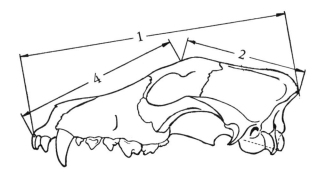

SKULL LATERAL ASPECT

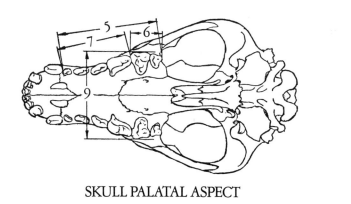

SKULL PALATAL ASPECT

P⁴

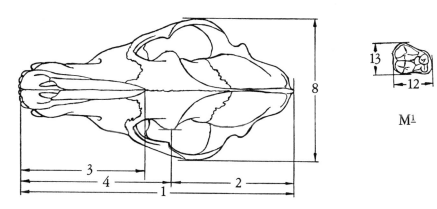

SKULL DORSAL ASPECT

Figure 8.1. Thirteen pertinent measurements selected as being diagnostic in separating wild from domestic species of *Canis* through mutivariate computer analysis. After von den Driesch (1976).

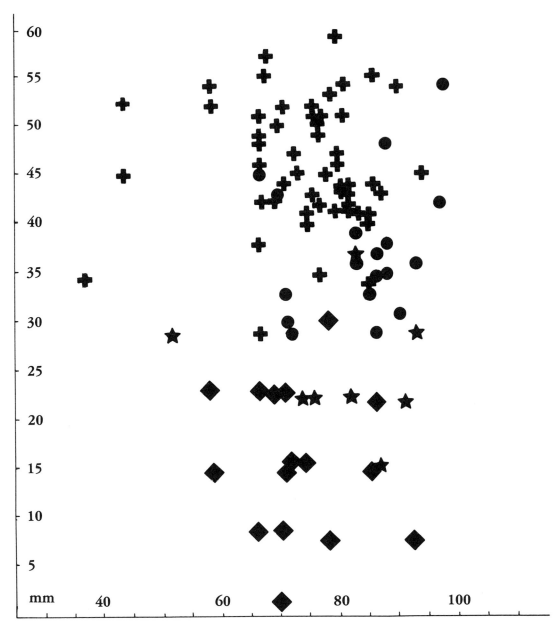

Figure 8.2. Results of a discriminant function analysis using thirteen variables for ● Alaskan Pleistocene wolves, ★ Alaskan Pleistocene short-faced wolves, ✚ modern wolves, and ◆ Eskimo dogs.

Premolar length	Measurement 7
Zygomatic width	Measurement 8
Molar width	Measurement 9
Width of P^4	Measurement 10
Length of P^4	Measurement 11
Width of M^1	Measurement 12
Length of M^1	Measurement 13

The purpose of performing a multivariate analysis on these measurements was to determine if it would be possible to isolate particular characteristics that would permit the separation of various species of *Canis* (Fig. 8.2). The initial study was exploratory in nature. There were numerous measurable characteristics for canid skulls that had been observed over the years, and there seemed to be no obvious reason why some should be more diagnostic than others for separating the species of *Canis* or why some should not be included as being of little use for this purpose. It was decided to utilize several variables, in combination, to attempt a separation of animals into similar groups.

I enlisted the aid of Dr. Harold Dibble, a computer analyst with the Department of Anthropology at the University of Arizona, who had conducted similar research on lithic assemblages and on human osteological variation. Dibble decided that a discriminant function contained in the Statistical Package for the Social Sciences, version 8.0 (Nie et al. 1975; Hull and Nie 1979) would fit the project as outlined. The program chosen, run on the Cyber 175 computer at the University of Arizona, enabled the multivariate comparison of two or more groups of individuals, with the ability to select those variables or combinations of variables that optimally discriminated between them. In addition to discrimination, the program contained an option for classifying cases into defined groups. This option was useful for assigning unknown specimens into particular groups of canid species and also, by re-classifying known individuals, for providing a check on the discrimination statistics.

The use and results of a computer analysis, as outlined above, is not the only approach to determining whether a canid is wild or domestic, but, rather, it is another important tool to be employed in the analysis of this intriguing problem. One conclusion that was verified by the computer program was that only multiple characteristics, based on more than fragmentary specimens, can allow for specific determinations of questionable canids.

LOCATIONS OF ILLUSTRATED SPECIMENS

REFERENCES

Allen, G. M.
    1920   Dogs of the American aborigines. Bulletin of the Museum of Comparative Zoology, Harvard University, vol. 63, no. 9, pp. 431–517. Cambridge, Massachusetts.
    1938   The mammals of China and Mongolia. Natural History of Central Asia, vol. 11, part 1, The American Museum of Natural History, 620 pages. New York.

An Zhimin
    1979a   Peiligang, Cishan he Yangshao: Shilun Zhongyuan Xinshiqi wenhua de yüanyüan jifazhan. Kaogu, no. 4, pp. 335–346. Beijing.
    1979b   Lüelun sanshinianlai woguo de Xinshiqi shidae kaogu. Kaogu, no. 5, pp. 393–403. Beijing, China.
    1981   Zhongguo de Xinshiqi shidai. Kaogu, no. 3, pp. 252–260. Beijing.

Anyang Archaeological Team, Institute of Archaeology, Chinese Academy of Social Sciences
    1979   1969–1977 Nian Yinxü Xiqu muzang fajüe baogao. Kaogu Xüebao, no. 1, pp. 27–146. Beijing.

Barker, B. C. W., and Macintosh, A.
    1979   The Dingo—a review. Archaeology and Physical Anthropology, in Oceania, vol. 14, no. 1, pp. 27–53.

Bayard, D. T.
    1972   Non Nok Tha: The 1968 excavation. Studies in Prehistoric Anthropology, University of Otago, vol. 4, pp. 15–31. Dunedin, New Zealand.

Beebe, B.
    1978   Two new Pleistocene mammal species from Beringia. American Quaternary Association, Abstracts of the Fifth Biennial Meeting, September 2–4, University of Alberta, p. 159. Edmonton, Canada.
    1980   A domestic dog (*Canis familiaris* L.) of probable Pleistocene age from Old Crow, Yukon Territory, Canada. Canadian Journal of Archaeology, no. 4, pp. 161–168. Toronto, Canada.

Behrensmeyer, A. K., and Hill, A. P.
  1980 Fossils in the making, vertebrate taphonomy and paleoecology. University of Chicago Press, 338 pages. Chicago.
Bellwood, P.
  1978 Man's conquest of the Pacific. Oxford University Press, 462 pages. New York.
Björk, P. R.
  1970 The Carnivora of the Hagerman local fauna (late Pliocene) of southwestern Idaho. Transactions of the American Philosophical Society, New Series, vol. 60, part 7, 54 pages. Philadelphia.
Bökönyi, S.
  1974 History of domestic mammals in Central and Eastern Europe. Akadémiai Kiado, 597 pages. Budapest, Hungary.
  1975 Vlasac: An early site of dog domestication. In: Clason, A. T., Archaeo-Zoological Studies. American Elsevier Publication, 477 pages. New York.
Boriskovsky, P. I.
  1968– Vietnam in primeval times. Published in 7 parts in Soviet Anthropology
  1971   and Archaeology, vol. 7, no. 2, pp. 14–32; vol. 7, no. 3, pp. 3–19; vol. 8, no. 1, pp. 70–95; vol. 8, no. 3, pp. 214–257; vol. 8, no. 4, pp. 355–366; vol. 9, no. 2, pp. 154–172; vol. 9, no. 3, pp. 226–264.
Brain, C. K.
  1981 The hunters or the hunted? University of Chicago Press, 365 pages. Chicago.
Brass, E.
  1904 Nutzbare Tiere Ostasiens Neudamn. Allen (1938) cited this reference in his bibliography but stated "not seen by author." I have not seen this reference nor have I been able to locate it through a library search.
Brown, B.
  1908 The Conard Fissure, a Pleistocene bone deposit in Northern Arkansas with descriptions of two new genera and twenty species of mammals. Memoirs of the American Museum of Natural History, vol. 9, part 4, pp. 155–208. New York.
Burleigh, R.; Clutton-Brock, J; Felder, P. J.; and Sieveking, G. de G.
  1977 A further consideration of Neolithic dogs with special reference to a skeleton from Grime's Graves (Norfolk) England. Journal of Archaeological Science, vol. 4, pp. 353–366. Dorset, England.
Canby, T. Y.
  1979 The search for the first Americans. National Geographic, vol. 156, no. 3, pp. 330–363. Washington, D.C.
Chang Kwang-Chih
  1977 The Archaeology of Ancient China. Third edition, Yale University Press, 535 pages. New Haven.
Chard, C.S.
  1974 Northeast Asia in prehistory. University of Wisconsin Press, 214 pages. Madison.

Clark, J. G. D.

    1954   Excavations at Star Carr: An early Mesolithic site at Seamer, near Scarborough, Yorkshire. Cambridge University Press, 200 pages. Cambridge, England.

    1972   Star Carr: A case study in bioarchaeology. Addison-Wesley Module in Anthropology. Cummings Publications Co., 42 pages. Menlo Park, California.

Clutton-Brock, J.

    1969   Carnivore remains from excavations of the Jericho Tell. In: The domestication and exploitation of plants and animals, edited by P. J. Ucko and G. W. Dimbleby, Aldine Publishing Co., pp. 337–345. Chicago.

Clutton-Brock, J.; Corbett, G. B.; and Hills, M.

    1976   A review of the family Canidae with a classification by numerical methods. Bulletin of the British Museum (Natural History), Zoology, vol. 29, no. 3, pp. 119–199. London.

Colbert, E. H.

    1950   The fossil vertebrates. In: Haury, E. W., The stratigraphy and archaeology of Ventana Cave, Arizona. The University of Arizona Press and the University of New Mexico Press, 599 pages. Tucson and Albuquerque.

    1955   Evolution of the vertebrates. John Wiley and Sons, Inc., 479 pages. New York.

    1973a   Further evidence concerning the presence of horse at Ventana Cave. The Kiva, Arizona Archaeological and Historical Society, vol. 39, no. 1, pp. 25–33. Tucson.

    1973b   Wandering lands and animals. E. P. Dutton and Co., Inc., 323 pages. New York.

Davis, S. J. M., and Valla, F. R.

    1978   Evidence for domestication of the dog 12,000 years ago in the Natufian of Israel. Nature, vol. 276, no. 5688, McMillan Journals Ltd., pp. 608–610. London.

Degerbol, M.

    1961   On a find of a preboreal domestic dog (*Canis familiaris*) from Star Carr, Yorkshire, with remarks on other mesolithic dogs. Prehistoric Society Proceedings, vol. 27, no. 3, pp. 35–49. London.

Dikov, N. N.

    1977   Arkheologicheskie Pamyatniki Kamchatki, Chukotki, i Verkhnei Kolymy. Aziya Na Styke s Amerikoi v Drevnosti, Izdatel Stvo Nauka, 390 pages. Moscow.

Emslie, S. D.

    1978   Dog burials from Mancos Canyon, Colorado. The Kiva, Arizona Archaeological and Historical Society, vol. 43, nos. 3–4, pp. 167–182. Tucson.

Epstein, H.

    1971   The origin of the domestic animals of Africa. Africana Publishing Corporation, 2 vols., 1292 pages. New York.

Ewer, R. F.

    1973   The carnivores. Cornell University Press, 494 pages. Ithaca, New York.

Fiennes, R.
    1976   The order of wolves. Bobbs-Merrill, 206 pages. New York.

Fine, M. D.
    1964   An abnormal $P_{\overline{2}}$ in *Canis* cf. *C. latrans* from the Hagerman Fauna of Idaho. Journal of Mammalogy, vol. 45, no. 3, pp. 483–485. Lawrence, Kansas.

Fox, M. W.
    1971   Behavior of wolves, dogs and related canids. Harper and Row, 220 pages. New York.
    1975   The wild canids. Van Nostrand Reinhold Company, 508 pages. New York.
    1978   The dog. Garland STPM Press, 296 pages. New York.

Galbreath, E. C.
    1938   Post-glacial fossil vertebrates from east-central Illinois. Geological Series of the Field Museum of Natural History, vol. 6, no. 20, pp. 303–313. Chicago.
    1947   Additions to the flora of the late Pleistocene deposits at Ashmore, Illinois. Transactions of the Kansas Academy of Sciences, vol. 50, no. 1, pp. 60–61. Lawrence, Kansas.

Gazin, C. L.
    1942   The late Cenozoic vertebrate fauna from the San Pedro Valley, Arizona. Proceedings of the U.S. National Museum, vol. 92, pp. 475–518. Washington, D.C.

Giddings, J. L.
    1964   The archaeology of Cape Denbigh. Brown University Press, 331 pages. Providence, R.I.

Gidley, J. W.
    1913   Preliminary report on a recently discovered Pleistocene cave deposit near Cumberland, Maryland. Proceedings of the U.S. National Museum, vol. 46, pp. 93–102. Washington, D.C.

Gollan, K.
    1977   (Personal communication.) Australian National University; Prehistory Department; RS Pac S, Box 4, P.O. 2600; Canberra, Australia.

Gorman, C. F., and Charoenwongsa
    1976   Ban Chiang: A mosaic of impressions from the first two years. Expedition, vol. 18, pp. 14–26.

Gray, J. E.
    1863   Notice of the Chanco Wolf (*Canis chanco*) from Chinese Tartary. Proceedings of the Zoological Society of London, London.

Green, M.
    1948   A new species of dog from the lower Pliocene in California. Bulletin of the Department of Geological Sciences, University of California, vol. 28, no. 4, pp. 81–90. Berkeley.

Guernsey, S. J. and Kidder, A. V.
    1921   Basketmaker caves of Northeastern Arizona. Papers of the Peabody Museum of Archaeology and Ethnology, Harvard University, vol. 8, no. 2, 121 pages. Cambridge, Mass.

Haag, W. G.
  1948   An osteometric analysis of some aboriginal dogs. Reports in Anthro-
         pology, vol. 7, no. 3, The University of Kentucky, 264 pages. Lexington.
Hall, E. R., and Kelson, K. R.
  1959   The mammals of North America. Ronald Press, 2 vols., 1083 pages. New
         York.
Hall, R. L.
  1978   Variability and speciation in canids and hominids. In: Hall, R. L., and
         Sharp, H. S., Wolf and Man. Academic Press, chap. 8, pp. 153–177. New
         York.
Hall, R. L., and H. S. Sharp
  1978   Wolf and man. Academic Press, 210 pages. New York.
Handan City Cultural Relics Management Team and the Handan Prefecture Cishan
    Archaeological Team Worker–Peasant Training Class
  1977   Hebei Cishan Xinshiqi yizhi shijüe, Kaogu, no. 6, pp. 361–372. Beijing.
Haury, E. W.
  1950   The stratigraphy and archaeology of Ventana Cave, Arizona. The Uni-
         versity of Arizona Press and the University of New Mexico Press, 599
         pages. Tucson and Albuquerque.
Hebei Province Cultural Relics Management Team and the Handan City Relics
    Preservation Station
  1981   Hebei Wu'an Cishan yizhi. Kaogu Xüebao, no. 3, pp. 303–338. Beijing.
Henan Provincial Museum; Yangzi Basin Planning Office; and Cultural Relics and
    Archaeology Team, Henan Division
  1972   Henan Xichuan Xiawanggang yizhi de shijüe. Wenwu, no. 10, pp. 6–19.
         Beijing.
Hibbard, C.
  1941   Mammals of the Rexroad fauna from the upper Pliocene of South-
         western Kansas. Transactions of the Kansas Academy of Sciences, vol.
         44, pp. 265–313. Lawrence, Kansas.
Hingham, C. F. W.; Kijingam, A.; and Manly, B. F. J.
  1980   An analysis of prehistoric canid remains from Thailand. Journal of
         Archaeological Science, vol. 7, pp. 149–165.
Hill, F. C.
  1972   A middle archaic dog burial in Illinois. Information Circular, Founda-
         tion for Illinois Archaeology, 8 pages. Evanston, Illinois.
Ho Ding-Ti
  1975   The cradle of the East. The Chinese University of Hong-Kong and the
         University of Chicago Press. 440 pages.
Hole, F.; Flannery, K. V.; and Neely, J. A.
  1969   Prehistory and human ecology of the Deh Luran Plain. Memoirs of the
         Museum of Anthropology, no. 1, University of Michigan, 438 pages. Ann
         Arbor.
Honacki, J. H.; Kinman, K. E.; and Koeppl, J. W.
  1982   Mammal species of the world. Allen Press, Inc., and the Association of
         Systematics Collections, 694 pages. Lawrence, Kansas.

Hopkins, D. M.
 1967   The Bering land bridge. Stanford University Press, 495 pages. Stanford.
Hu Chang-kang and Qi Tao
 1978   Gongwangling Pleistocene mammalian fauna of Lantian, Shaanxi. Pal-
        aeontologia Sinica, Whole Series no. 155, New Series C no. 21, 64 pages.
        Beijing.
Hull, C. H., and Nie, N. H.
 1979   SPSS Update: New procedures and facilities for releases 7 and 8.
        McGraw Hill, 172 pages. New York.
Institute of Archaeology; Chinese Academy of Social Sciences; and the Banpo,
        Xian, Shaanxi Museum
 1963   Xian Banpo-Yuanzhi minzu gongshe juluo yizhi, Zhongguo tianye
        kaogu baogaoji, Kaoguxüe zhuankan, Dingzhong shisihao, Wenwu
        Chubanshe, 320 pages. Beijing.
Jia Lanpo and Zhang Zhenbiao
 1977   Henan Xichuan xian Xiawanggang yizhizhong de dongwuqün. Wenwu,
        no. 6, pp. 41–49. Beijing.
Johnston, C. S.
 1938   Preliminary report of the vertebrate type locality of Cita Canyon and the
        description of an ancestral coyote. American Journal of Science, series 5,
        vol. 35, pp. 383–390. New Haven.
Jones, F. W.
 1921   The status of the Dingo. Transactions of the Royal Society of South
        Australia, vol. 50.
Kahlke, H. D., and Chow Ben-Shun
 1961   A summary of stratigraphical and paleontological observations in the
        lower layers of Choukoutien, Locality 1 and on the chronological posi-
        tion of the site (in Chinese). Vertebrata PalAsiatica, no. 3, pp. 212–240.
        Beijing.
Kaifeng Prefecture Cultural Relics Management Team and Xinzheng County Cul-
        tural Relics Management Team
 1978   Henan Xinzhang Peiligang Xinshiqishidai yizhi. Kaogu, no. 2, pp.
        73–79. Beijing.
Kaifeng Prefecture Cultural Relics Management Team; Xinzheng County Cultural
        Relics Management Team; and Zhengzhou University History Depart-
        ment, Archaeology Section
 1979   Peiligang yizhi yijiuqibanian fajüe baogao. Kaogu, no. 3, pp. 197–203.
        Beijing.
Keightly, D. N.
 1978   Sources of Shang history. University of California Press, 281 pages.
        Berkeley.
Kelley, J. E.
 1975   Zooarchaeological analysis at Antelope House: Behavioral inferences
        from distributional data. The Kiva, Arizona Archaeological and Histor-
        ical Society, vol. 41, no. 1, pp. 81–85. Tucson.
Kurtén, B.
 1959   The carnivora of the Palestine caves. Acta Zoologica Fennica, vol. 107, 74
        pages. Helsinki.

1972   The Age of Mammals. Columbia University Press, 250 pages. New York.

1974   A history of coyote-like dogs (Canidae, Mammalia). Acta Zoologica Fennica, vol. 140, 38 pages. Helsinki.

Kurtén, B., and Anderson, E.

1980   Pleistocene mammals of North America. Columbia University Press, 442 pages. New York.

Laboratory of the Institute of Archaeology, Chinese Academy of Social Sciences.

1974   Fangshexing tansu ceding niandai baogao (san). Kaogu, no. 5, pp. 333–338. Beijing.

1979   Fangshexing tansu ceding niandai baogao (liu). Kaogu, no. 1, pp. 89–94, 96. Beijing.

1980   Fangshexing tansu ceding niandai baogao (qi). Kaogu, no. 4, pp. 372–377. Beijing.

1981   Fangshexing tansu ceding niandai baogao (ba). Kaogu, no. 4, pp. 363–369. Beijing.

Larsen, H., and Rainey, F.

1948   Ipiutak and the Arctic Whale Hunting Culture. Papers of the American Museum of Natural History, vol. 42, 276 pages. New York.

Lawrence, B.

1956   Appendix E: Cave fauna. In: Field, H., An anthropological reconnaissance in the Near East, 1950. Papers of the Peabody Museum of Archaeology and Ethnology, Harvard University, vol. 48, no. 2, pp. 80–81. Cambridge, Mass.

1967   Early domestic dogs. Sonderdruck aus S. F. Säugetierkunde, Band 32, Heft 1, pp. 44–59. Hamburg.

1968   Antiquity of large dogs in North America. Tebiwa, vol. 11, no. 2, Idaho State University Museum, pp. 43–49. Pocatello, Idaho.

Lawrence, B., and Bossert, W. H.

1967   Multiple character analysis of *Canis lupus, latrans* and *familiaris* with a discussion of the relationship of *Canis niger.* American Zoologist, vol. 7, pp. 223–232. Utica, N.Y.

1969   The cranial evidence for hybridization in New England *Canis.* Breviora, no. 330, Museum of Comparative Zoology, Harvard University, 13 pages. Cambridge, Mass.

Lawrence B., and Reed, C. A.

In press   The dogs of Jarmo.

Lehmer, D. J.

1973   (Personal communication.) Dana College; Blair, Nebraska.

Leidy, J.

1885   Notice of remains of extinct vertebrata from the valley of the Niobrara river. Proceedings of the Academy of Natural Sciences of Philadelphia, p. 21. Philadelphia.

Li Youheng and Han Defen

1959   Shaanxi Xi'an Banpo Xinshiqi shidai yizhi zhong zhi shoulei guge. Gu Jizhui Dongwu yü Gu Renlei, vol. 1, no. 4, pp. 173–186. Beijing. Reprinted 1963.

1963   Banpo Xinshiqi shidai yizhi zhong zhi shoulei guge. Appendix 2, pp. 255–269. In Xian Banpo—Yuanzhi minzu gongshe juluo yizhi, edited

Li Youheng and Han Defen *(Continued)*
   by the Institute of Archaeology, Chinese Academy of Social Sciences and
   the Banpo Zian, Shaanxi Museum, Zhongguo tianye kaogu baoguoji,
   Kaoguxüe zhuankan, Dingzhong shisihao. Wenwu Chubanshe, Beijing.

Linnaeus, C.
   1758   Systema naturae. 10th ed., 824 pages. Uppsala, Sweden.

López, B. H.
   1978   Of wolves and men. Charles Scribner's Sons, 309 pages. New York.

McMillan, R. B.
   1970   Early canid burial from Western Ozark highland. Science, vol. 167, no.
          3922, pp. 1246–1247. Washington, D.C.

Martin, R. A.
   1974   Fossil mammals from the Coleman II A fauna, Sumter County. In:
          Pleistocene mammals of Florida, edited by S. D. Webb, University of
          Florida Press, pp. 35–99. Gainesville, Florida.

Martin, R. A., and Webb, S. D.
   1974   Late Pleistocene mammals from the Devil's Den fauna, Levy County. In:
          Pleistocene mammals of Florida, edited by S. D. Webb, University of
          Florida Press, pp. 114–145. Gainesville, Florida.

Matthew, W. D.
   1918   Contributions to the Snake Creek fauna. Article 7, Bulletin of the Amer-
          ican Museum of Natural History, vol. 28, pp. 189–190. New York.

   1924   Third contribution to the Snake Creek fauna. Bulletin of the American
          Museum of Natural History, vol. 50, pp. 59–210. New York.

   1930   The phylogeny of dogs. Journal of Mammalogy, vol. 11, no. 2, pp.
          117–138. Lawrence, Kansas.

Meadow, R.
   1983   Vertebrate faunal remains from Hasanlu 10 at Hajji Firuz: The Neolithic
          settlement. In: Voiat, M. M., ed., Hansanlu excavation reports, vol. I,
          University Museum Monograph 50. The University Museum, 430 pages.
          Philadelphia.

Mech, L. D.
   1970   The wolf. Natural History Press, 384 pages. Garden City, New York.

   1974   *Canis lupus.* Mammalian Species, no. 37. American Society of Mam-
          malogists, American Museum of Natural History, 6 pages. New York.

Medvedev, G. I.
   1969   Results of the investigation of the Mesolithic in the stratified settlement
          of Ust'-Belaia, 1957–1969. Arctic Anthropology, vol. 6, no. 1, pp. 61–69.
          Madison, Wisconsin.

Mellaart, J.
   1964   A Neolithic City in Turkey. Scientific American Reprint, no. 620, W. H.
          Freeman and Co., 11 pages. San Francisco.

Merriam, J. C.
   1906   Carnivora from the tertiary formations of the John Day region. Univer-
          sity of California Publications, Bulletin of the Department of Geology,
          vol. 5, no. 1, 64 pages. Berkeley.

   1910   New mammals from Rancho La Brea. University of California, Bulletin
          of the Department of Geology, vol. 5, no. 25, pp. 391–395. Berkeley.

Mertens, H.
1936 Der Hund aus dem Senkenberg. Moor, ein Begleiter des Urs. Nat. u. Volk, vol. 66, pp. 506–510.

Mivart, St. George
1890 Dogs, jackals, wolves, and foxes: A monograph of the Canidae. London.

Mochanov, I. U. A.
1967 Bel'Kachinskaia Neoliticheskaia kul'tura. Materialy IV Soveschchaniia po Okhrane Pirody Yakutii, pp. 51–56. Yakutsk.

Nguyen, Phuc Long
1975 Les nouvelles recherches archéologiques au Vietnam. Arts Asiatiques, vol. 31.

Nie, N. H.; Hull, C. H.; Jenkins, J. G.; Steinbrenner, K.; and Brent, D. H.
1975 Statistical package for the social sciences. Second edition, McGraw-Hill Publishers. New York.

Novikov, G. A.
1962 Carnivorous mammals of the fauna of the U.S.S.R.: Keys to the fauna of the U.S.S.R.. Zoological Institute of the Academy of Sciences of the U.S.S.R., no. 62. Translation of the 1956 original published by the Israel Program for Scientific Translations, 284 pages. Jerusalem.

Nowak, M. R.
1979 North American quaternary *Canis*. Monograph of the Museum of Natural History, University of Kansas, no. 6, 154 pages. Lawrence, Kansas.

Okladnikov, A. P.
1955 Neolit i Bronzovyy Vek uribaikaliya Chast' 3-Glazkovskoye vvemya, Matcrialy i Issledovaniya po Arkheologii SSSR, no. 43. Moscow.

Olsen, J. W.
1980 A zooarchaeological analysis of vertebrate faunal remains from the Grasshopper Pueblo, Arizona. Ph.D. dissertation in Anthropology, University of California, Berkeley. University Microfilms, 369 pages. Ann Arbor.

Olsen. S. J.
1970 Two pre-columbian dog burials from Georgia. Bulletin of the Georgia Academy of Science, vol. 28, pp. 69–72. Athens, Georgia.

Olsen, S. J., and Olsen, J. W.
1977 The Chinese wolf ancestor of New World dogs. Science, vol. 197, no. 4303, pp. 533–535. Washington, D.C.

Olsen, S. J.; Olsen, J. W.; and Qi Guo-qin
1980 Domestic dogs from the Neolithic of China. Explorers Journal, vol. 58, no. 4, pp. 165–167. New York.

Osborne, H. F.
1931 Lower Pleistocene age of Peking Man. Natural History Magazine, American Museum Expeditions and Notes, vol. 31, no. 4, American Museum of Natural History, p. 446. New York.

Patte, E.
1932 Le kjökkenmödding Neolithique de Da But et ses sépultures. Bulletin du Service Géologique de l'Indochine, vol. 19, fascicle 3.

Pei, W. C.

    1934   The carnivora from Locality 1 of Choukoutien. Palaeontologia Sinica, Series C, vol. 8, Fascicle 1, Geological Survey of China, 45 pages. Beijing.

    1936   On the mammalian remains from Locality 3 at Choukoutien. Palaeontologia Sinica, Series C, vol. 7, Fascicle 5, Geological Survey of China, 108 pages. Beijing.

Péwé, T. L.

    1975a   Quaternary geology of Alaska. Geological Survey Professional Paper 835, U.S. Government Printing Office, 145 pages. Washington, D.C.

    1975b   Quaternary stratigraphic nomenclature in unglaciated central Alaska. Geological Survey Professional Paper 62, U.S. Government Printing Office, 32 pages. Washington, D.C.

Pidoplichko, I. G.

    1969   Late Paleolithic dwellings of mammoth bones in the Ukraine. Institute of Zoology of the Ukrainian Academy of Sciences, Naukova Dumka, 162 pages. Kiev.

Powers, W. R.

    1973   Paleolithic man in northeast Asia. Arctic Anthropology, vol. 10, no. 2, 106 pages. Madison, Wisconsin.

Qi Guo qin

    1977   Fujian Minhou Tanshishan Xinshiqi shidai yizhi zhong chutu de shougu. Gu Jizhui Dongwu yü Gu Renlei, vol. 15, no. 4, pp. 301–306. Beijing.

Radinsky, L.

    1973   Evolution of the canid brain. In: Riss, W. Brian, Behavior and evolution. S. Karger Publishers, pp. 169–202. Basel, Switzerland.

Rainey, F. G.

    1939   Archaeology in central Alaska. Anthropology Papers of the American Museum of Natural History, vol. 36, part 4, pp. 351–405. New York.

Reed, C. A.

    1959   Animal domestication in the prehistoric Near East. Science, vol. 131, no. 3389, pp. 1629–1639. Washington, D.C.

    1960   A review of the archaeological evidence on animal domestication in the prehistoric Near East. In: R. J. Braidwood and B. Howe, Prehistoric investigations in Iraqi Kurdistan. Studies in Ancient Oriental Civilization, no. 31, The Oriental Institute of the University of Chicago Press, 184 pages. Chicago.

    1961   Osteological evidence for prehistoric domestication in Southwestern Asia. Sonderdruck aus "Zeitschrift für Tierzüchtung und Züchtungsbiologie" Band 76, Heft 1, pp. 31–38. Hamburg.

    1969   The pattern of animal domestication in the prehistoric Near East. In: The domestication and exploitation of plants and animals, edited by P. J. Ucko and G. W. Dimbleby, Aldine Publishing Co., pp. 361–380. Chicago. (This entry also published in 1970 by Gerald Duckworth and Company, Ltd., London.)

Repenning, C. A.
1967   Palearctic-Nearctic mammalian dispersal in the Late Cenozoic. In: D. M. Hopkins, The Bering land bridge, Stanford University Press, 495 pages. Stanford.

Rine, J. Z.
1965   The world of dogs. Doubleday and Company, Inc., 331 pages. Garden City, New York.

Sadek-Kooros, H.
1972   Primitive bone fracturing: A method of research. American Antiquity, vol. 37, no. 3, pp. 369–382. Washington, D.C.

Schenk, E. T., and McMasters, J. H.
1956   Procedures in taxonomy. Stanford University Press, 119 pages. Stanford.

Schlosser, M.
1903   Die fossilen saügetierre Chinas nebst einer odontographic der recenten antilopen. Abhandl. K. Bayerlschen Akad. Wiss. Math. Phys. Cl., vol. 22, part 1, 221 pages. Berlin.

1927   Weitere bemerkungen über fossile carnivoran aus China. Palaeontologia Sinica, vol. 4, Fascicle 4, Geological Survey of China, pp. 8–10. Beijing.

Scott, J. P.
1950   The social behavior of dogs and wolves: An illustration of socio-biological systematics. Annals of the New York Academy of Sciences, vol. 51, article 6, pp. 1001–1122. Albany, New York.

Shetelig, H., and Falk, H.
1937   Scandanavian archaeology. Oxford University Press, 458 pages. Oxford.

Simpson, G. G.
1945   The principles of classification and a classification of mammals. Bulletin of the American Museum of Natural History, vol. 85, 350 pages. New York.

Singh, P.
1974   Neolithic cultures of western Asia. Seminar Press, 240 pages. New York.

Skinner, M. F.
1942   The fauna of Papago Springs Cave, Arizona. Bulletin of the American Museum of Natural History, vol. 80, article 6, pp. 143–220. New York.

Smith, C. H.
1839   Dogs. The naturalist's library series; W. J. Bart, editor, Mammalia; vol. IX, 267 pages. Edinburgh.

Solheim, W. G. II
1972   An earlier agricultural revolution. Scientific American, vol. 226, no. 4, pp. 34–41.

Sommarström, B.
1956   Archaeological research in the Edsen-Gol region of Inner Mongolia. Part I, Statens Etnografiska Museum, Stockholm.

Speth, J. D., and Parry, W. J.
1980   Late prehistoric bison procurement in southeastern New Mexico. The 1978 Season at the Garnsey Site (LA-18399). Museum of Anthropology, the University of Michigan Technical Reports on Archaeology, contribution 7, 369 pages. Ann Arbor.

Stock, C. M.
    1938   A coyote-like wolf jaw from the Rancho La Brea Pleistocene. Bulletin of the Southern California Academy of Sciences, vol. 37, part 2, pp. 49–51. Los Angeles.

Stroganov, S. U.
    1969   Carnivorous mammals of Siberia. Academy of Sciences of the USSR, Siberian Branch. Translated from the 1962 original by the Israel Program for Scientific Translations, 521 pages. Jerusalem.

Struever, S., and Holton, F. A.
    1979   Koster. Anchor Press/Doubleday, 281 pages. Garden City, New York.

Tedford, R.
    1978   History of dogs and cats. Ralston Purina Co., 10 pages. St. Louis.

Teilhard de Chardin, P., and Pei, W. C.
    1941   The fossil mammals from locality 13 of Choukoutien. Palaeontologia Sinica, New Series C no. 11, Whole Series no. 126, Geological Survey of China, 106 pages. Beijing.

Teilhard de Chardin, P., and Piveteau, J.
    1930   Les mammiferes fossiles de Nihowan (China). Annales de Paleontology, vol. 19, pp. 88–89. Paris.

Tolstoy, P.
    1958   The archaeology of the Lena Basin and its New World relationships. American Antiquity, vol. 23, no. 4, part L, pp. 397–418. Washington, D.C.

Triestman, J. M.
    1972   The prehistory of China. Natural History Press, American Museum of Natural History, Doubleday and Company, Inc., 156 pages. Garden City, New York.

Tringham, R.
    1969   Animal domestication in the Neolithic cultures of the southwest part of European U.S.S.R. In: P. J. Ucko and G. W. Dimbleby. The domestication and exploitation of plants and animals, 581 pages. Aldine Publishing Co., pp. 381–392. Chicago.

Turnbull, P. F., and Reed, C. A.
    1974   The fauna from the terminal Pleistocene of Palegawra Cave. Fieldiana, Anthropology, vol. 63, no. 3, Field Museum of Natural History, pp. 81–145. Chicago.

Vanderhoof, V. L., and Gregory, J. T.
    1940   A review of the genus Aelurodon. University of California Publication, Bulletin of the Department of Geological Sciences, vol. 25, no. 3, pp. 143–164. Berkeley.

Van Gelder, R. G.
    1978   A review of canid classification. American Museum Novitates, no. 2646, American Museum of Natural History, 10 pages. New York.

von den Driesch, A.
    1976   A guide to the measurement of animal bones from archaeological sites, Peabody Museum Bulletins, no. 1, Peabody Museum of Archaeology and Ethnology, copyright 1976 by the President and Fellows of Harvard College, 135 pages. Cambridge, Mass.

Webb, S. D.
    1974   Pleistocene mammals of Florida. University of Florida Press, 270 pages. Gainesville, Florida.

West, F. H.
    1980   Archaeology of N.E. Asia. The Quarterly Review of Archaeology, vol. 1. Williamstown, Massachusetts.

Xia Nai
    1977   Tan-shisi ceding niandai he Zhongguo shiqian kaoguxüe. Kaogu, no. 4, pp. 217–232. Beijing.

Young, S. P., and Goldman, E. A.
    1944   The wolves of North America. The American Wildlife Institute, 636 pages. Washington, D.C.

Zdansky, O.
    1924   Jungtertiäre carnivorean Chinas. Palaeontologia Sinica, Series C, vol. 2, fascicle 7, Geological Survey of China, pp. 10–14. Beijing.
    1927   Weitere bemerkungen über fossile carnivoran aus China. Palaeontologia Sinica, Series C, vol. 4, fascicle 4, Geological Survey of China, pp. 8–10. Beijing.

Zeuner, F. E.
    1958   Dog and cat in the Neolithic of Jericho. Palestine Exploration Quarterly, pp. 52–55.
    1963   A history of domesticated animals. Hutchinson of London, 560 pages. London.

Zhejiang Provincial Cultural Relics Management Team and Zhejiang Provincial Museum.
    1976   Hemudu faxian yuanzhi shehui zhongyao yizhi. Wenwu, no. 8, pp. 6–12. Beijing.
    1978   Hemudu yizhi diyiqi fajüe baogao. Kaogu Xüebao, no. 7, pp. 39–94. Beijing.

Zhejiang Provincial Museum, Natural History Section
    1978   Hemudu yizhi dongzhiwu yicun de jianding yanjiu. Kaogu Xüebao, no. 1, pp. 95–107. Beijing.

Zhou Benshun
    1980   (Personal communication.) Institute of Archaeology; Beijing.
    1981   Hebei Wu'an Cishan yizhi de dongwu guhai. Kaogu Xüebao, no. 3, pp. 339–347. Beijing.

INDEX